# A Peerie Peek at da Past

# A Peerie Peek at da Past

## Agnes Hobbin

Published by
The Shetland Times Ltd.,
Lerwick.
2000

A Peerie Peek at da Past



ISBN 1 898852 69 3

First published by The Shetland Times Ltd., 2000.



British Library Cataloguing-in-Publication Data
A catalogue record for this book is available from the British Library.

Printed and published by
The Shetland Times Ltd.,
Prince Alfred Street,
Lerwick, Shetland, ZE1 0EP, UK.

# CONTENTS

# ILLUSTRATIONS

# ACKNOWLEDGEMENTS

I wish to thank my son, George Hobbin, who's idea it was to publish this book. Thanks also to my niece, Maisie Colligan, who read it over and agreed with him, and to Sandra Sutherland who typed it for me.

# Prologue

ON the 7th day of July, 1921, I was born in a box bed, in the ben end of a croft house at Geosetter, Bigton. I was my mother's seventh child and my father's twelfth. Father's first wife, who came from Burra Isle, had died from tuberculosis. His second wife, my mother, came from Quarff. Her christian name was Ursula and her maiden name was Davidson. She was a real kindly person who loved her step-children just like her own. I feel sure they also loved her. My father's name was George John Goudie. In those days it was the custom to call children after a relative. I was given the name Agnes, after an aunt in Quarff.

I am led to believe that the first Gaudie (subsequently Goudie) to settle at Geosetter was my grandfather, Laurance Gaudie. He married in Dunrossness to Andriana Cheyne of Exnaboe in 1854. How they came to reside in Geosetter I don't know, as Laurance was born in Rerwick, Bigton.

Father was a crofter fisherman with a share in a fishing boat called the *Maggie Paterson*. She usually sailed from Levenwick, on the east side of Shetland, occasionally coming in to Vedrick on the west side, just below Geosetter. When this happened father would take my siblings and me in a small boat so that we could go aboard the *Maggie Paterson*. We did enjoy going down the ladder into the ship's hold.

The croft at Geosetter was a good fertile croft, able to sustain six cattle, plus two working horses. It also carried a souming for about twenty-six sheep in the hill common grazing. To help sustain us we kept hens and one or two pigs. The crops we grew consisted of bere, corn, potatoes, cabbage and neeps. My parents also had a share in a croft at Clavel. On this croft only Shetland cross Cheviot sheep were kept, the sale of their lambs bringing in a little welcome money. One good thing about this croft was its sandy soil, so new potatoes were ready for eating three or four weeks before those on the Geosetter croft.

# CHAPTER 1
# Looking Back to Croft Life

WHEN one reaches three score years and ten, plus five, it is not unusual to look back. After all, if looking forward, one could not envisage a very long look.

In my case looking back takes me to a croft house in the south end of Shetland. Nothing, in my opinion, can compare with the comfort of a croft house in the evening. In my mind's eye I can see it now, the lamp-light falling on father at one side of the open peat fire, mending nets, and mother at the other, knitting. As my brothers grew older they took a turn at the net mending and we girls (my sisters and I), learned to knit at an early age.

Being born into a large family meant one was never lonely, always someone ready to talk or play. When not busy, father would read aloud from *The Shetland Times* or the *News of the World*. We did enjoy his readings from *The Shetland Times*, particularly the weekly readings from the Shetland book *Lowrie*. Many, many years later when my husband asked what I would like for my birthday, I requested a copy of this book. My request was granted and I have the book to this day. The *News of the World* was not so interesting, but father insisted we must know what was going on in the world. My mother shared the *People's Friend* with a neighbour and we children used to love reading "Willie Waddel and Wag". A number of *Christian Heralds* arrived from time to time, passed on from someone, and on Sundays mother would read out the sermons. A home-made bookcase, painted a lovely coral colour, hung on the wall behind father's chair. We younger ones thought the books it housed very dry reading; the Bible, a dictionary, *Pilgrim's Progress*, some sea stories and an assortment of Sunday School prizes, mostly religious.

Down below the bookcase hung a home-made canvas bag affair, ornamented with tassels. It had two compartments, one for tidying up papers and the other for any unsolicited adverts. I hasten to add there was very little unwanted mail in those days. All furniture in the house was wooden and home-made; beds, chairs and tables, all made by father. The croft house itself was built from stone, erected by father and his brother. The inner end of the but end floor was wood and the outer flag stones.

At a very early age we learned we were lucky to have security of tenure. This had come about by an Act of Parliament in 1886, I think called the Crofter's Reform. No longer could the lairds turn out families to make room for sheep. Unfortunately, one now has to worry in case one's home is sold to pay for one's keep in old age, and also to wonder how long one will be able to heat one's home, with VAT on electricity etc. There is something to be said for the mossy peat, if one had the strength to work it.

The croft provided most of our food. We were well fed, even if our diet was monotonous. Breakfast was beremeal porridge while our home-grown beremeal lasted, then it was oatmeal porridge, the exception being a few weeks in winter. This came about when only one cow was milking. We had boiled mutton, onion and bannocks for breakfast. Later on in life we had porridge made with what was called "flake meal", I suppose the forerunner of porridge oats. Weekday dinner at 1pm, except when at school, was either salt herring or salt fish and boiled potatoes. Sunday dinner was always meat.

Once a year a fat cow (mert) was collected by the butcher, killed, cut up and brought back to be shared out with uncle and aunts next door. The only pay the butcher asked was to retain the beast's skin. He also made for us lots of saucermeat seasoned with sugar and saltpetre, so it could be preserved for a long time. In my home it was stored in a large crock and opened in the voar, when more calories were required to cope with extra work.

My mother made the cow's puddings, some with currants and some with oatmeal. They were hung in white flour bags in the rafters and used alternately with eggs for tea.

Cooking was done on the open fire, with the kettle, or pot, hung in the crook. Bannocks were baked on the brand iron. Coals were drawn out from the fire, placed under the brand iron and the bannocks baked on top. I now, in my old age, try to re-enact this by baking my scones under the grill, but while good, they cannot compare with the childhood bannocks.

Once in a while mother made a cake in the Dutch oven. This oven was placed near the fire, with coals underneath and coals on top, and this made as good a cake as one baked in a modern electric cooker.

Father was not only a crofter, but a fisherman as well. At the end of the herring season he brought home food for the winter. This consisted of two bolls of flour and one of oatmeal, two barrels of herring (one for our aunts and one for us), and a barrel of rough salt. He also brought home for each of his young children (he had fathered twelve), a pair of hobnailed boots. Fishermen in those days did not pay off with a great deal of money, but some boats were more lucky than others. One year he told us of a fisherman who had enough money to buy boots for his children but not enough for the laces. Fortunately the shopkeeper gave him the laces free. My father had a share in his boat, the *Maggie Paterson*, so paid off with a bit more money. One other thing he brought home for his younger children was a garment called a combination. This affair covered the upper limits as well as one's trunk, and was always of a dusky pink. It had little holes at the back of the neck that, when counted, determined the age it could fit. The last big item was always two drums of paraffin — winter fuel for the lamps. The smallest item was a big bag of sweets, usually liquorice allsorts. I suppose they were the cheapest, but to this day I often spend a little of my retirement pension on a quarter of the same, and suddenly I feel young again.

When the herring season was in full swing, father would bring home, at the weekend, a bag of fresh herring. They were filleted and spread out on a board to dry in the sun. We had sun in those days! When dry they were seasoned with a shake of salt and pepper and roasted on the brand iron. This was a great treat, as was fresh olicks (ling) or haddocks, when weather and time allowed the male members of our family to go fishing.

Round about October or November, a fat sheep, or perhaps two, were killed. This was a nice change from salt herring or salt fish, as was reestit mutton. Reestit mutton was mutton pickled in salt for a few weeks, then hung up about the peat fire to dry. Very few people from the south appear to like the particular flavour of this meat. One doctor I knew, when asked if he was busy replied; "Indeed I am; it is no wonder considering the amount of mummified carcasses the Shetland people eat."

We did have grocery vans, but only essential things like tea and sugar were ordered from them. A typical order from the van would be something like this:

2oz of brown tobacco (father smoked)
2lbs of sugar
1/2lb of tea
1lb of baking soda
1/4lb cream of tartar
1 bar of preserving soap
1 bar of Lifebuoy soap
1 bag of blueing (used to whiten white sheets)

Occasionally a few other items would be added to the order, like a pound of dried fruit — for the Dutch oven cake, and a tin of pepper. We did not require to buy table salt. Mother crushed some rough salt between two pieces of paper, with a bottle.

All summer we did a lot of churning of milk. We had butter, kirn milk, butter milk and blaand for baking. Any butter we did not require at the time was salted down for the winter. If this did happen to run out, mother ordered from Orkney. When I was old enough to write, I was given the task of requesting the butter. We seldom had bought jam, my mother made about 40lbs of rhubarb jam — the fruit being home-grown or from our aunt's in Quarff.

My siblings and I were brought up to honour and respect our elders and, when very young, to be seen and not heard, except to answer a question. Unless ill, we attended the church every Sunday. On Saturday our shoes were brushed, the water barrels scrubbed out and filled with clean water. We children were all bathed and our hair washed in a zinc bath in front of the open fire. Sunday was, as much as possible, a day of rest, except for feeding the humans and animals. Dinner at 1pm was more elaborate than on a weekday — meat, cabbage and turnip, boiled with potatoes, and sometimes a pudding to follow. After dinner we used to go for a walk, or sit out on the grass and make daisy chains. At 4pm we had home-made pancakes and then all walked to church. My elder sisters and brothers all sang in the choir and we younger ones sat like mice in the pew, with mother and father. My elder brother rang the church bell twice per Sunday for the services. He also lit the lamps in the wintertime and saw to the cleaning of the church.

Father was in some ways very Victorian; we had to do exactly as he said. He disapproved of card games ("the Devil's Books") and, until we left home, we were not allowed to attend dances. The exception was a dance to finish up the night school (teachers would be present) or a wedding. Consequently, if we did go to either of those, we did not know how to dance. Lots of our peer group were the same, so the M.C. would announce the "Grand Old Duke of York" or the "Jolly Miller" and then we would join in.

In other ways our father was very modern. No child of his was ever to have a hand laid on him or her. Once a year he invited our school teachers to our home for "eight o'clocks". He made it quite clear that no strap was ever to be laid on his children. One teacher, who later became his daughter-in-law, dared to ask if he did not believe if "one spared the rod, one spoiled the child". His answer; "Any teacher who could not control a class, without the strap, should not be in post."

We, as a family, found our father very stern and forbidding. I suppose he had suffered a lot. His first wife had died young, and very shortly after, their last baby had also died. He also lost a son from drowning. He slipped and fell into the water while climbing rocks above the sea.

Only once did any of us dare to play a trick on father. When not at the herring fishing he used to like a nap in the afternoon. One day one of my brothers placed a mousetrap in his bed just where his hand would connect when he laid back the bedclothes. It did connect, right over his thumb. He came through looking stern and angry. I thought he would lose his temper and thrash my brother. No such thing. He turned to the culprit and said, "I know it is you. Your punishment is no dinner for three days." Actually he did not go without dinner. Mother put his aside and after we had eaten and father had gone out, he got his. Looking back I think it was probably arranged between father and mother that he would be fed when father went out.

In those days it must have been very difficult to punish a child without recource to a belt or a slipper. After all, one did not get pocket money as such — very occasionally we might get a penny. As one shared one's bedroom with a sibling or two, one could not be isolated.

Mother was the sort of person who could not bear anyone to be hurt, hungry or cold. She always saw the best in everyone. She had a saying thus; "There is so much bad in the best of us, so much good in the worst of us, that it ill behoves the best of us to say any ill of the worst of us." I never did know the origin of the saying, but from our mother we learned a great deal of tolerance.

When very young, I used to be sent next door, to our aunt's house, to borrow reading material for father. He never could get enough to read. Once when there, I heard one aunt remark to the other, "our house always drops soot". On returning home I asked my mother what this meant, as I was not aware of any soot dropping from the chimney. Mother said, "It means a child is present who might tell things out of our home." She then explained that what happened in one's own home was the business of the people in that home, and no-one else, and one must learn never to repeat things. She finished up by saying, "learn young, learn fair".

One of my half-sisters had been a short time at college to learn dressmaking and embroidery, so it was she who made most of our clothes. Every year mother ordered a parcel of remnants, costing about 10/-. From this parcel she got enough material for dresses for two of my sisters and myself, and enough thick material for a skirt for herself. Every week she knitted a jumper or cardigan with a bit of Fair Isle or embroidery. She had a cousin in Bathgate who sold them for her. She received 10/- for a cardigan without sleeves and 12/6 for one with sleeves. At an early age we children made the pockets — even the few minutes this saved mother was a great help. Somehow or other there always seemed to be enough money for Sunday clothes. In those days that included hats and we did enjoy getting dressed up on a Sunday.

Our jumpers were all home-made by members of our family, or by ourselves when we were old enough. Once when my dressmaking half-sister was knitting thick jumpers for sale, she had some wool left over. From this wool she knitted two caps, each with a lovely tassel on the crown. One was white and the other was a sort of wheat colour. I was disappointed as a sister a little older than myself got the white one, and I got the other. I was only about four years of age, but I took the white cap outside, placed it on a stone, and was looking around for a sharp stone to put holes in it when the knitter of the caps came out. What was I doing? I confessed, "I wanted the white cap as the other is a dirty colour." She laughed and said I did not need the white cap. I had rosy cheeks and bright auburn hair. My sister, she explained, was a blonde with very little colour in her cheeks, so she required a lighter cap. Not a cross word spoken, I took the cap in and laid it carefully where it had come from.

Shortly after this the owner of the white cap was admitted to hospital for a tonsillectomy. I don't remember much about her going, but I do recall her coming home. An aunt carried her up to our house and she had been given a ball and a doll. Looking at them I felt that I could part with my tonsils anytime if this was the reward. Then one of my half-brothers spoke and said that he would have something for me when he came home next weekend. True to his word, he brought for me a little black rubber doll. I don't suppose it cost much, but oh how I loved it and, when old and battered, did not want to part with it. Then this same half-brother said it was time that "Glaxo" — that was the name I gave him — had a rest. He took the little rubber doll and a spade in one hand, and myself in the other, and we went up to a bit of hilly ground. There he dug a hole and planted "Glaxo" in it. Next he said a few comforting words, and hoped he would rest in peace. I had noticed that the grave was under a nice clump of heather bells and made up my mind to recover "Glaxo" when no-one was looking. Unfortunately, I underestimated the grave digger, for when I went back nothing was there.

Next weekend this same half-brother arrived home with two boxes, about the size of shoe boxes. One was plain white and the other floral. I was asked to choose one for myself and needless to say I took the floral box. Later on in life when I heard the expression, "Don't judge the chocolates by the box," I was very much in agreement. The plain box housed a doll with a lovely china head, my box contained a doll with a soft head. So a sister two years older than myself got the better doll. I knew it was fair, I had chosen, and I truly did love my doll, and it had a long life.

When I was a little older this same sister and myself were given almost identical dolls, all done up in white lace and cotton wool. Some time previously we had heard the story of "Snow White and the Seven Dwarfs" and I was just about to say that I would call my doll Snow White when my sister forestalled me. I protested, but one of my half-sisters said, "Lets choose a name for your doll that no-one else has." After some deliberation we decided on "Puffy-Waggey". "Puffy" because she was puffed up with cotton wool and "Waggey" because her headgear swung from side to side when she was moved. We did so like this half-sister; she gave us scraps of material and trimmings to play with and we had great fun with tins of buttons and buckles. In our pretend games the buttons became pupils and the buckles teachers.

Home-made toys were also a diversion. Mother could carve a kettle from a potato, stick a button on top for a lid and a match for the spout. She also made chairs and tables from cardboard, paper roses from tissue paper and little boats from stronger paper. Once I wished out loud that I had a proper draughtboard — ours was just made from cardboard. At this remark father looked up from the paper he was reading and said, "Find me a proper piece of wood and I will carve you one." A quick look around the croft did not reveal anything suitable so I set off for the beach at Vedrick. I must have been about ten because I was allowed to go on my own. While there I found a piece of wood the sea had thrown up and it must have been out of the sea for some time as it was nice and dry. When presented to father, out came the ruler and he said, "I can't believe it. It is the correct size and the proper kind of wood for carving. No excuse, I will have to get on with it." He did get on with it and just with a chisel and a knife he carved me a lovely board.

Apart from a good football, my brothers' toys were also home-made. Whistles, boats and miniature ships made from wood, and a big piece of wood over a barrel for a seesaw. One of my brothers was very inventive and built a cycle out of old pieces. When I started school I got a lift on the wooden crossbar. He also built a little house from stones with turf and straw for the roof. It even had a chimney that could vent and we did enjoy it so much. I was very young at the time and can't remember what happened to it. Perhaps father had it flattened in case of fire.

At Christmas, Santa Claus came with a Santa sock. This usually contained a cardboard house to build, a trumpet affair which made a loud noise, a little puzzle and in the toe of the sock a few tiny sweets. Sometimes he was extra generous and left an apple and an orange tied to the sock. On Christmas Day mother always prepared a nice tea, the highlight of this being an iced cake. This was a gift from a cousin of my mother's who was a cook to a countess.

My first social event was to see a baby that was born at Smirdale, the croft next to ours. Mother dressed my sister and myself in our Sunday best clothes and we set out. We went over the Geosetter Burn and up the other side to Smirdale. There were stone steps at our side but just stones at the other so we had to pick our feet as we had on our best shoes. We were well received by an elderly woman who, I think, must have been the granny. First of all she boiled the tea kettle in the crook over the fire, then pulled out coals from the fire, put tea and hot water in a lovely brown teapot, and set it on the coals. Next she made pancakes. Last of all she set down on the table the most delicate china. The cups and saucers were white with lovely red roses and the china so thin it almost looked blue. I started to count the cups and saucers and the people, and the elderly lady said, "The bairn can count, what are you counting for?" I replied, "To see if I will get one of the lovely cups to drink from." My mother protested but the lady said, "If the bairn wants to drink out of a nice cup, so shall she drink." I was then sat down on a little home-made chair with a creepie (stool) in front of me, on which was placed pancakes and, best of all, a beautiful cup and saucer. At home I had a mug, my very own, with a picture of a teddy bear on the side, but it could not compare with having a cup and saucer and feeling grown up. After the tea drinking ceremony, the baby was brought out from the but end. She was wrapped in a lovely white lace Shetland shawl, which was turned down and under her feet. Everybody admired the baby, but when she was about to be taken ben again, the granny-like lady said, "My bairn, do looks worried, what is it?" When I replied, "Does the baby only have one foot?", she loosened the shawl and I saw two lovely white feet. Looking back I can't help admiring a grown-up who could give so much consideration to a little girl like myself.

At this time in my story, a little word about the burn. We crossed at the shallow end, but it did have some very deep holes. As children we were informed that a big fish called a pilder lived in the big holes and, if we went near it would get angry, jump out and run after us. Needless to say we never went near the big holes till we were old enough to take care, and knew why we were informed about the pilder. As we grew older, we did enjoy the burn with the mayflowers, honeysuckle and ferns.

Last year I had a visit from a nephew from Wick. He too was brought up in Geosetter and we started to reminisce about the pilder. One day when his elder brother, two cousins and himself were playing rather near the burn they heard a roar. The others ran crying, "The pilder, the pilder." He, however, thought there was something familiar about the sound, turned to have a look and there was his father's head appearing above the burn.

We were brought up never to fear the dark. If we were lucky, we might hear the little people (trows) playing the fiddles at night. If extra lucky, we might catch a glimpse of one of them. They were supposed to be very tiny, wore black trousers, red jackets and each had a lovely little toorie on his head. Needless to say we never did catch a glimpse.

One thing I did enjoy in the dark was accompanying mother to the byre when she went to milk the cows. She carried a lantern and, when the weather was frosty, I almost felt like Good King Wenceslas's page, as I walked in her footsteps. At this time I was allowed to feed the cat or cats in the byre. They always got warm milk straight from the cow. Father did not allow cats or dogs in the house. He was very health conscious and thought they might pass on disease to humans.

Another thing we were never frightened of was thunder and

lightning. We were to look out of the window and report if fork lightning, as that was supposed to clear the air quicker.

Not long after the birth of the baby at Smirdale, the family sold up and went to live in Walls. Father attended the sale and bought a cring of lambs. He said to my mother, "I had to buy something, they have been such good neighbours." Even at my early age, of perhaps four, I could see he felt vexed they were leaving.

About this time one of my half-brothers decided to get married. Now the work began with a quarrying of stones for the foundation and walls for his home to be. It was just a living-room and bedroom to be built on to our house, but it was to have its own door.

I was the only one of our family not old enough to go to school, so one day my half-brother invited me to come and see how his house was getting on. I refused. Mother had just made me a pair of slippers and I said they would get wet. The slippers were made out of an old black jumper and a piece of cardboard between two pieces of material for the sole. The laces were bright coloured pieces of yarn, twisted and tied in over the top. Last of all, a little flower was embroidered on the front of each slipper. Not to be daunted, he carried me and showed me where everything was to go. The bedroom would be small, but as he earned more money he would build a bigger one. The living room would be big, like the old fashioned but end, and they would have a real stove. What luxury! In no time at all, with help from father and three brothers, the building was complete.

Plans then went ahead for the marriage and house reception. The meal was set in our uncle and aunt's house, which had been built on to our other side. I don't remember much about the meal, but I do about the musical evening afterwards in the new home. My new sister-in-law had an organ and someone played the fiddle and I was amazed to see mother and father dancing what I now know to be the Shetland Reel. There were games for us children as well, things like "Farmer in his Den" and "Musical Chairs". When it was getting late, mother wanted me to go to bed. I refused but did agree to lie down in my bed, just for a little rest. My older sister and myself both had white dresses, hers trimmed with blue and mine with rose coloured flowers. I did not wish to take off my dress, but finally was persuaded it would crush, so I agreed on condition that I could keep on my new petticoat that was trimmed with lace.

When I awoke it was morning and I had a tantrum. It was not so much that the party was over, it was the feeling I had been tricked. I adored my mother and thought she was above reproach, how could she tell a lie like this. Later on in life I learned this was what one would call a white lie. Just then the bridegroom came in from next door. What was the matter, why was I crying? After explanations he brought forth a big bag of reading sweets. I was invited to sit down with him and he would read them to me. "Oh what does this one say?" "You are lovely", and the next,

"Meet me tonight." He then started to sing "Meet me tonight in the cow shed, meet me tonight all alone," and soon I was laughing and enjoying the sweets. It was a case of what the elderly Shetlanders used to say, "It does not take much to amuse children." I think it would take more in this day and age, what with TV and children carrying around their personal stereos and CD players.

Soon a big day arrived, I was starting school. Five children went out from our home, two boys and three girls. I was allowed to sit with an elder sister in the big school room for a week or two, until I got used to the new environment. A copy book was placed in front of me and I was asked to make cup-hooks. I refused but my sister took the book from me and, under the desk, made them for me. The teacher was suitably impressed with what she thought was my own work.

Something happened after I had been a few days at school. The older girls had rigged up a swing in the shed. Plenty of girls were willing to push me and I clambered to go higher and higher, till all of a sudden I fell out onto the concrete floor. I don't remember about that bit, or the fact that I was carried home unconscious, but history relates that when I woke up I was given a dose of castor oil.

All too soon, it was nearly holiday time and the photographer arrived. My four siblings and myself had our photo taken together. It came out well. I have a copy and treasure it to this day.

After the holiday only four of us went back to school. The elder of my two brothers went to the Anderson High, then known as the Institute. He got a good bursary, otherwise I don't suppose my parents could have afforded to let him go. I missed him and also my lift to school on the crossbar of his bike, but fortunately he came home every weekend.

My first year at school was very ordinary. We had little reading books, learned how to write a little and played with plasticine. At the intervals we played games — "Rounds", "Croonie" and "Puss-Puss come to my corner" and such like. Sometimes we quarrelled, mostly in fun. We would chase after someone calling, "I will get you," and laughing so much it was impossible to catch up. Just about the end of that year, my brother came home from the Institute very ill and soon most of our family were also ill. Looking back, I think it must have been scarlet fever or scarlatina. Mother got rheumatic fever and a bed had to be put up in the but end for her. She was too ill to sleep in bed with my father. He got up in the middle of the night to put peat on the fire, to keep the room at a constant temperature. He also hung a sail over the door to prevent draughts. Mother was unable to turn in bed, so father and my thirteen year old sister had to help her. This same sister had to get an exemption from school to cook and clean for us all. No Social Services in those days. My elder half-sister was teaching in Sandness and my other one was next door looking after my uncle, his wife and two elderly aunts. My two half-brothers were at sea and my eldest brother was living next door, doing the croft work for my uncle so we had to manage as best we could. It was not easy, four children under fourteen years of age and one of fourteen. Fortunately neither my father nor my elder sister got the infection, but the rest of us did. I became very ill indeed. My eardrums burst and hydrogen-peroxide had to be installed, twice per day, to clear out the puss. No antibiotics in those days, so when it was realised I had a kidney infection, the only treatment was diet. This was very scanty, not even milk, I had to drink buttermilk. I was a great trial if I could not eat and drink what I wished; sometimes I refused to eat or drink at all.

One night I had a strange dream which has stayed with me all my life. I thought I was in a big black box surrounded by wonderful exotic flowers. I kept trying to pick them but was never successful. Later on in life they were just like flowers on the first funeral wreath I saw. Much later in life still, as a student nurse in Edinburgh Royal Infirmary, I kept a close eye on any very ill patient who appeared to be plucking flowers with his or her fingers.

Dr Laing came frequently and tested my urine. Any day he did not come father took my temperature. In those days it was taken in the mouth and I hated it. After what seemed a very long time, the doctor decided I could have a little more food. Still no meat, but I could have lentil soup. My thirteen year old sister cooked the soup and I did eat a little, but the onions felt raw. I picked them out and arranged them round the side of my plate, like a decoration. My poor sister, what a life she had. I wonder that she did not throttle me.

One day when father was looking out the window, he said, "What next! Here comes the Compulsory Officer." This was a new one on me. I knew about ships officers. My two half-brothers hoped in time to become captains, and I knew about police officers, from the *News of the World*, but what on earth was this? Very soon I found out. He wanted to know why I had been so long off school. Father informed him I was improving a little but was in no way able to walk a mile to school. He said he believed in education but would rather have an illiterate live daughter than an educated dead one. He finished by saying, "If you do not believe me, you have my permission to ask Doctor Laing." The officer replied he did most certainly believe him, it was just his duty to come. Our dinner had all been cleared away, but he was given two boiled eggs, bannocks and tea. After this he departed, with a sad look at me, and expressed the hope that I might soon feel better.

Well I did improve — very slowly. With hindsight, I feel it was the strict diet that pulled me through. My mother improved also slowly. The day she was able to get out of bed seemed like a miracle. We had missed her cheery smile and her way of sorting our ills about the house.

Time marched on and one day when the better weather came, it was decided I should start school again. A letter accompanied me, requesting I be allowed to go to the toilet whenever I wished, and this was granted. In a way I had sort of looked forward to going back to school, but the reality was not good. After such a long time without lessons, I was well behind. The teacher had a bad time with me, as I did not take kindly to any reproof. She called me lazy, which put my back up. I did not feel lazy, I just felt tired. Looking back I feel sure I was anaemic. After all, my diet had been mostly carbohydrate for some time. One day the teacher informed me if ever I had all my spellings correct she would give me something. That day I went home and learned all my spelling from the spelling book for the next day, plus those that might be asked from the reading lesson. The reading lesson was "Jason and the Golden Fleece". When the spellings were corrected the teacher said, "Guess who goes to the top of the class, all spelling right." She then went to a cupboard and brought out six lovely propelling pencils. They were shaped like golf clubs and were in different colours. How I loved them and they lasted for ages. We never did become friends, but fortunately after a year or two she left to be nearer her family in the south.

We then got a Shetland girl for a teacher and I liked her from the start, and I began also to like going to school. Previously I often pretended to be ill, just to stay home; I only had to give a little cough and I was grounded for a few days.

When I was about ten my uncle died. His widow came in to tell us, and my father went immediately with her to attend to arrangements. As our doctor had seen him recently, he did not require to come, just to be informed. We as children, were taken in, one by one, to see him lying in his coffin — this was to be our way of paying our last respects. Mother explained that it was just his body lying there, his spirit had gone up to Heaven. That night I almost had nightmares, wondering if his spirit might return, and what form it might take.

About a year after this, my dressmaking half-sister got married. What excitement led up to it. My elder sister was bridesmaid and my other sister and myself flower girls. As usual, our dresses were all home-made, different material but all in pink, and we flower girls had white shoes and ankle socks. The reception was in our aunts' house; sandwiches, fancy buns and bride's cake, a little music afterwards and more games. The groom was a lay preacher, who later became a Reverend so no dancing was allowed. The actual marriage ceremony was in our local church, open to all who wished to attend, and all friends and neighbours came.

Shortly after this, our living-room (but end) got a face lift. A new floor was installed. Out went the chain and crook and in went a range. My fifteen year old brother made a new dresser out of driftwood. This dresser was given to me later on in life and I, in turn, have given it to my son who has it to this day. One thing that vexed me very much was the replacement of two lovely pictures with wedding photos. The pictures had hung over the mantelpiece for as long as I could remember. One was of a young boy being questioned by some officials and had the caption below; "When did you last see your father?" The royal velvet of their jackets looked so real, one could almost stroke them. The other was of a lady and a child in a beautiful park, plucking flowers near a stream. Looking at this picture I almost felt I could run my fingers through the water.

Life settled down after this, in a humdrum sort of way. As we grew older we had to work. In the holidays we had to hoe tatties, single neeps, cole hay and help in getting the peats home from the hill. In the harvest time we had to work with the corn. As the youngest, I was given the job of making bands for the sheaves of corn. These were made by tying sticks of corn at the grain end. I hated this as my hands got so cold. The lifting of the potatoes was also a big chore, the older ones delling with Shetland spades and we younger ones glenning (gathering) up the potatoes. What a sore back one had at night.

One day when I was about eleven years of age I came home from school to see a doctor's car at the road below our house. I knew someone was not well and it transpired that one of my aunts had had a stroke. Now it was all hands to the pumps. Father laid down the rota. The night was divided, one sat with her from 11pm to 4am, then they woke someone else and they sat till 8am. My job was to go in after tea-time and read the Bible to her. Fortunately she preferred the New Testament which made more interesting reading. Her favourite passage from the Bible was, "In my Father's house are many mansions, were it not so, I would have told you, I go to prepare a place for you, that where I am there you may be also." Although her speech was not as it had usually been, I was able to understand her. She said she was not afraid to die, she was looking forward to seeing her fiancé, who was lost at sea just before they were due to marry. She had had a dream that she would be seeing him on the shortest day. I was also supposed to put my hand under her pillow, raise her head and give her drinks from time to time, so her mouth could be kept moist. There seemed to be no tissues in those days, or else we just did not have them. However, little pieces of clean cotton were at the bedside, to rub her mouth when required. We had no electricity so warming of the bed was done with a thing called a pig, an earthenware hot water bottle. Unfortunately this did not seem to help aunt's feet, so I was instructed to knit a pair of white Shetland wool bedsocks. This I did with help from mother turning the heel. Aunt only lived about three weeks so she did not have long to enjoy them. She died on the shortest day. As was the custom we had to pay our last respects. We were taken one by one to see her lying in her coffin. She looked composed and all strain seemed to have left her face. We children were kept off school till after the funeral, and then life resumed as normal.

Looking back and thinking of care in hospital, and in the community, I don't think anything could compare with care in the old-fashioned extended family. Light nourishing food and plenty of fluids given. Two persons, usually an adult and an older child, turned the patient carefully, washed her and changed her night-dress. Her hair was then brushed and combed and a hot protected pig put to her feet. After this she relaxed and often had a short nap.

The next bit of trouble was the arrival home of a brother feeling ill. At this time he was employed as a cook on the fishing boat *Maggie Paterson*. Father, now retired, still had a share in this boat and one by one as his sons left school they served a year or two as cook. At first my brother just had a vague feeling of malaise, slight headache and flu like symptoms. After a day or two, however, he had to take to his bed. My mother at that time was also confined to bed with acute rheumatism, something she got quite often since she had rheumatic fever. The thermometer was brought out and after a day or two, when the temperature started to creep up, the doctor was sent for. He arrived in the morning when the temperature was lower and we were pleased to hear it was just a flu type illness and he would soon be better. Not so, the doctor was sent for again and father informed him he wanted something done, as his son's temperature was a little higher every night. The doctor was not pleased, he said our thermometer must be wrong but finally decided to transfer my brother to hospital, as he said it could be inflammation of his appendix. An older sister travelled in the ambulance with him and stayed in Lerwick overnight. We had no phone so waited anxiously for her homecoming the next day, hoping the operation would be over and all would be well. Fortunately the surgeon was not happy that the symptoms suggested appendicitis, and transferred him to the Isolation Hospital where paratyphoid fever was diagnosed.

Now began the great clean up, the straw mattress and pillows were burned, all sheets and blankets washed with disinfectant and the wooden bed washed and scrubbed, also with disinfectant. Lastly the room was fumigated with sulphur sticks. These were always available in our home as they were used to whiten shawls. Lastly all cups and plates my brother had used were boiled for twenty minutes. When the Sanitary Inspector arrived there was nothing left for him to do but test the water supply. No fault was found in it; it ran freely from a spring.

The infection was traced to a well. The *Maggie Paterson* put in somewhere for supplies and my brother went to a well for water. It was a hot day and he drank some water on the way back to the boat. The other members of the crew only drank it in tea, the water having been boiled, so did not get the infection.

Strange to say no-one else in our household got it, but we children were excluded from school for about three weeks. My brother seemed to be in hospital for a very long time, and when he did come home he had lost his hair. Fortunately it grew in again and this time curly.

Life resumed again, the daily grind, school all weekdays and helping at home in the evenings. On Saturdays, if not required to work on the croft, we were expected to clean in the house. The wooden chairs had to be scrubbed and the floors had not only to be washed but polished as well, as we had advanced to linoleum. The furniture also had to be washed and polished. I always had to "black lead" the range, as my sister two years older than me refused to take a turn. She said when she grew up all the housework she wished to do would be to arrange flowers. This did not happen. She had to work for a living like everybody else.

We had some good times as well, the best being when we were allowed to visit our aunts in Quarff. They made us very welcome and we were allowed to play cards and enjoy the company of cousins our own age. I was once there at Christmas time and we went guising through all the houses in Quarff, and when we came back an uncle-in-law played the fiddle.

Another event in our lives was the marriage of my elder half-sister. She arrived home one weekend with a very tall gentleman who was the headmaster at the school she was teaching in. My youngest sister and I were a bit frightened of him, he looked so large, and had very big feet. He then came out of his pocket with a lovely new penny each, which although just copper, looked like spun gold. Lastly out of the pocket came a big bag of caramels. This was accompanied with a lovely smile which lit up his face, and we really felt he was quite handsome. I seem to remember the expression "handsome is as handsome does". The reception was a quiet affair but we children did enjoy the bride's cake which seemed to last for ages.

The next event in our lives was a sad one; one of my half-brothers died. He was the one who built a but end and bedroom for his bride, and hoped to build more. Sadly this never happened. He died young leaving a widow and small daughter.

However life had to go on. All my siblings were working and only one brother, working at the fishing came home at the weekend. It was a rather lonely time. I was thirteen and in my last year at school. I never did finish that year. My brother arrived home one weekend with measles and I was immediately excluded from school. I got the infection and by the time I was well enough to go back, it was almost holiday time, and as my birthday fell during the holidays, I was never back.

Now began a new chapter in my life, it was crofting work from morning till night. During the Voar and Hairst we usually had paid help, but ordinary croft work we did ourselves. Among other things, I sledged home the peats from the hill with a horse that had to have blinders (blinkers) on as he was liable to take off if startled. I had the good sense to give him a pat from time to time so he could feel secure, and we got on fine. The worst times were when my mother had to take to her bed with acute rheumatism. This meant I had to milk three cows, put them out to pasture and "flit" them twice per day. Father loosened them from the stalls and put their tethers on, but could not do much more as he was bothered with arthritis. There was also the pig and the hens to feed and fresh water to be carried in. Cooking was very basic but still had to be done. Washing of very dirty clothes had to be done with the scrubbing board. For my work I received my board and keep, but was expected to knit for my clothes. When I became old enough, father insured me as an agricultural worker. This was because having been quite often ill and off school as a child, he wished some security for me if I became really ill again. In those days there was no National Health Service and all doctor's bills had to be paid.

When I became sixteen I asked to be allowed to apply for a paid job. By the time the croft work was done there was very little time to knit, and I wanted more new clothes. My request was refused but I was informed that when we brought out enough chickens for our own use I could bring out more and sell for money for myself. The agreement was that I got the eggs free, but provided the paraffin for the incubator. Apart from this I had to sprinkle the eggs with water and turn them twice per day, read the thermometer and adjust the lamp, to keep the incubator at the right heat for the eggs to hatch. This was no problem, as I had been doing it for our own chickens. A lady, whom my parents called a "hen wife", came from Lerwick to teach me to distinguish a female chicken (new-born) from a male one. I quite enjoyed this work, the incubator took 50 eggs and the brooder could easily deal with that amount, and best of all I made money.

I was able to buy from J.D. Williams' catalogue one green costume, two blouses — one green and one white, and new sandals. The costume cost 35/- and the blouses 7/6 each. I can't remember the cost of the sandals. I was also able to take money from my knitting kitty to buy a hat. How I did love that hat; it was green with a white ribbon and was supposed to be in the Ginger Roger's style.

When I was about seventeen, my parents decided to give up croft work. Actually a brother of mine, living next door, had been doing the croft work for some time. Father was at that time seventy-one and mother sixty. They did, however, keep the hens, about seventy of them, and one pig. We kept the hens free from infection by putting into their drinking water a substance called permanganate of potash. They also had proper drinking pails which allowed a little water to come when they drank, so no contamination. We shipped the eggs to a firm called Goodwin in Glasgow. This was more lucrative than doing a barter system at the grocery van.

We were now into 1939 and the news was not good. Mr Chamberlain came back from Germany carrying his umbrella and we had a short breathing space.

Now my parents decided, as they had given up croft work, I must be trained for something so I could provide for myself when they were no longer there. I now had two half-sister married and settled, and two sisters training to be nurses, one in Edinburgh Royal and one in Dunfermline. Nursing was thought to be out for me; I had much less education than my sisters and was not considered to be strong. It was decided that I should go to Craibstone College to learn cookery. An application was filled in for a bursary and I got a good reference from my ex-schoolteacher. Then fate took a hand. A communication arrived from the matron of the Gilbert Bain Hospital saying she had had two of my sisters for probationers before they went south to train. She now had a vacancy if I wished to take it up. It was decided I could go, but was to leave when my bursary came through.

On the very day that war was declared, I started nursing in the Old Gilbert Bain Hospital. Duty started at 6.15am and lasted till about 9.30pm, with two hours off during the day. One of my last jobs at night was to carry down, with another probationer, the dirty bins and burn the contents in the furnace. It was hot work, also dirty and smelly, particularly if there had been an amputation during the day.

Two little incidents occurred when I was there. It was the practice at that time to have bedpan rounds; it was considered almost unsocial to wish to pass urine at any other time. We usually carried four empty bedpans, one on top of each other, separated by a piece of rubber. Once when I was doing exactly this, and proceeding down the corridor about to enter the female ward, I glanced up at a young apprentice electrician repairing a ceiling light. Our eyes met and he jumped down and gave me a cuddle and a kiss. Unfortunately matron happened to come round the corner at that precise moment. Did I get a telling off? I was a disgrace to my uniform. I protested I had nothing to do with the young man giving me a kiss. Definitely I did not wish to be kissed carrying four bedpans. Later on he apologised saying he never thought he would get me into trouble. I was young enough to laugh, but hoped it wouldn't happen again. I had no wish to be sent home to feed hens and a pig.

The next incident was quite different and happened when I was on night duty. There were only two of us on duty, a trained nurse and myself. Being the junior, I was the person who went down to warm up the food left for us by the cook. To get to the kitchen I had to pass the mortuary door and just at that time it started to open and a vision in a large white gown appeared. I was just about to scream when the vision said, "Don't scream, it's only me", and it turned out to be one of my fellow workers in a long evening dress. She was actually sleeping in the hospital at that time and explained if one wanted to be late, one opened the mortuary door from both sides and closed them again after one got in. At that time in our career we were expected to be in at 10pm.

On that same night duty, matron came to waken me one afternoon to tell me that my half-brother's ship, the *Giralda,* had been sunk off Orkney by enemy action. No-one had survived. She said I could get compassionate leave. I could go home and my job would be kept open for my return. This was big blow; it was he who bought me the "Glaxo" doll, and many other presents. He also sent me postcards from foreign places, including the fjords in Norway.

When I arrived home there was only me to comfort my parents. Their grief was great. He had been promised the command of a ship on his next voyage and now his life was over. In the absence of my siblings I did my best. No-one else could get leave; my brothers were on the high seas — one in the Royal Navy, one in the Merchant Navy — and my sisters both training to be nurses in the south.

I had been home only a few days when my mother fell and broke an arm and attended hospital to have it set. Now my parents did not want me to go back to nursing. My bursary for Craibstone had come through but I did not want to go there. Fate again took a hand. Mother went into hospital to have her plaster off and when she came back she turned to father and said, "The surgeon wants Agnes to come back immediately. He says she is going to make a good nurse." Father said very little, but later on he said he was going to Lerwick the next day. As he went to Lerwick very seldom, I knew he had something in mind, but even I did not guess what it was. Well it turned out he had ordered a washing machine. It duly arrived, a wooden affair, rather like a big kirn for churning milk. We had no electricity so it had to be turned by hand. This was not difficult, a nice change from the scrubbing board. Now I was informed that I could go back to work, they would manage. Two sons and three daughters would be coming home on holiday from time to time, so life was not all black.

Back to nursing I went, no more croft work for me at Geosetter, except for brief holidays. A new way of life was before me, but that is another story. This one had ended.

# CHAPTER 2
# My Early Nursing Days

NURSE training for me started in Edinburgh Royal Infirmary in December 1940, after having spent just over a year in the Gilbert Bain Hospital, Lerwick as a probationer. Age for entry to Edinburgh Royal Infirmary was dropped to nineteen in 1940 because of the war. When one of my sisters wished to train there she had to wait until she was twenty-one.

I was grateful to be accepted as I had no secondary education. However, the matron of the Gilbert Bain Hospital gave me a reference, as did the Rev. R.N.T. Anderson, the Church of Scotland minister in the church I attended. It was the practice to have an entrance exam if one left school at fourteen years of age. Because of the long journey from Shetland in wartime, it was decided I could have my exam on the day I was due to start training.

My parents were a bit worried about the journey in wartime, especially as I had never been out of Shetland. They arranged for a fireman on the north boat to take me ashore, and a half-sister was to travel up from Dollar and meet me outside the security bit of the harbour. This took some thinking out, as I was not allowed to say when the boat was sailing. It was decided that when I knew, I would send a telegram saying, "I am well, Agnes". This I did, but did not bargain for a sub-postmistress at Blairingone putting on her own interpretation. She thought if I was well there was no hurry to deliver the telegram and had the boat not been delayed, I would not have been met. Later on in life when I became a sub-postmistress (a position I held for eighteen years) I got the telegrams delivered as quickly as possible.

The day I set out to get the boat for Aberdeen was a Wednesday, a day off in Lerwick, so no bus. I duly went down to the roadside carrying my suitcase to thumb a lift. Bigton was twenty-two miles from Lerwick. Petrol was rationed so there were few private cars on the road. Thinking about it now, I remember what an aunt said to my sister and myself later on in life. My sister had just started driving and wondered what we would do if we had a puncture. My aunt said, "Lasses, just stand at the side of the road and look helpless and beautiful and you will soon get help!" Well at the time I was looking for a lift in December 1940, I was certainly not beautiful! I must have looked forlorn and helpless though, for in no time at all a driver picked me up and delivered me to the steamer's store in Lerwick where I was able to leave my suitcase. It was almost post office shutting time before I knew we were going to sail, but I did manage to get my telegram off.

In those days females were all in one cabin and gentlemen in the other. That night the cabin was occupied only by two ladies and myself. They cried all night, saying if the Germans did not sink us, the bad weather would. Never having been on this journey before, I was not aware that it was a very bad night. Indeed I did not feel frightened and slept, in between being sick and listening to the crying. In the morning the stewardess came to apologise for not looking to us during the night but she had been so sick she could not keep her feet!

Next the fireman arrived, asking if I was alright. He informed me that the weather was so bad we had lost a lifeboat. He carried my suitcase, escorted me ashore, and delivered me to my half-sister. As it was now lunchtime, instead of 7am — our expected arrival time, we had something to eat in a cafe, before getting the train for Dunfermline and the bus for Blairingone. How I did enjoy the train journey to Dunfermline. The beautiful trees and gardens flashing past, the wonderful buildings and the bustle and commotion of people getting off and on at railway stations. Next was the bus journey, and it was even nicer than the train. It was slower and I could get a better view of everything as we passed. All too soon we arrived at my half-sisters home, and I spent Thursday, Friday, Saturday and Sunday there. My brother-in-law was the minister for the parish, so they lived in the Manse. It had once been a laird's house so the rooms were many and large. There was even a butler's pantry, but of course no butler, not even a maid to make one feel one should be living in such a big house. Time passed quickly. One day we had a bus run and Sunday was the day for the church.

Early on Monday morning I set off for the train to arrive in Edinburgh at the appointed time, I think 12 midday. Fifteen other girls arrived about the same time. I asked them about the entrance exam. It was just arithmetic and English, so I thought no problem, but in any case the powers that be forgot about it, and I did not remind them.

We all had our names taken, a small speech of welcome and then lunch. After lunch we were medically examined, then measured for uniform. Next we were escorted to our sleeping quarters in the Red Home, used for the most junior nurses. We each had a room to ourselves, plenty of room for the few small garments we possessed, the only drawback was the place was terribly cold and the blankets on the bed were worn and thin. No Shetland sheep's wool there, to be sent to Hunters and spun into cosy wool blankets. In my opinion, there is no better conductor of heat than wool, and no better wool than that from Shetland sheep.

As new girls, we were supposed to have seven weeks in Preliminary Training School, but as we would be there over Christmas, we were told we would be nine weeks. This was a big blow, as we did not get any pay in P.T.S. and had to pay a fee of £5.50 to be admitted. This was towards meeting state examination fees, and was non-returnable.

As probationers, we were to receive £20 for the first year, £25 in the second, £30 in the third and in the fourth £40. We were to be provided throughout with board, lodgings, laundry and the regulation amount of uniform material. Holidays were four weeks per year.

During our stay in P.T.S. we received instruction in domestic science, sick room cookery, bandaging and such practical nursing details as could be practised outside the wards. We also had lectures in elementary anatomy, physiology and hygiene. We all had a turn at various household duties, the worst being pantry duty. On this duty we started at 5.30am which was not much fun as we were on summer-time all winter. The wooden pantry shelves had to be scrubbed each day and quite honestly were beginning to wear thin. Three of us would be on breakfast duty, one set the table, one made butter-pats with the old fashioned scotch hands, and the third served. Dishes all had to be washed, then examined by the sister in charge, who always managed to find a tea leaf in a cup. We concluded that she carried the tea leaf around with her, in order to give a lecture on how, if one could leave a tea leaf in a cup, one might leave a swab in a patient's abdomen. The knives were the old fashioned kind that had to be scoured with bath brick. Milky cocoa was served at 11am, also prepared by ourselves, and the dishes washed up. Lunch at 1pm was in the main dining room, a big austere room with one framed text on the wall. It read; "To hands that work and eyes that see, give wisdom's heavenly lore, that whole and sick, and weak and strong, may serve thee evermore". Afternoon tea was also made by ourselves, and dishes washed up. Supper was in the main dining room again, usually mealy pudding, stovies or herring.

Our bedrooms were examined by sister tutor every morning, to make sure we had opened our windows — at the top and bottom — and laid down our bed to air.

Out and in among, we had lectures and were instructed to handle bones, to help us in our anatomy. We did not like our sister tutor, and as far as I could gather, no P.T.S. had ever liked her.

Early in February we started work in the wards. It was a big change for most of the girls, but not for me. Having been a year in the Gilbert Bain Hospital, I settled down to ward life as if I had never been away. I made many friends among my fellow workers, although we did not have much time to socialise.

A bell wakened us at 6.15am and we had to be in the dining room for roll call, and morning tea at 6.30am, having previously laid down our bed to air. Then back to our rooms to make our beds and report to our respective wards by 7.15am.

My first ward was male medical, no big deal, but the ward sister was a bit of a tartar. We used to love when she had her half-day and we would try to quicken up our work so we could have a little extra time in the afternoon. No such luck, she had eyes like a hawk. The first time I tried it, she said, "Nurse Goudie, you seem very far forward today. I will give you two extra bed baths to do this afternoon."

I only had one real misanter while there. Most of the patients had coughs, and all had sputum mugs. They had to be taken out, cleaned and brought back on a tray, along with a bottle of disinfectant. As one set the mug on the locker, one added a little disinfectant. Unfortunately for me the bottle was top heavy, toppled over and crashed to the floor, in the middle of the chief's round, with all the students round about. I expected a big row, but no such thing, sister even gave me a shy smile and I realised that underneath that grave exterior there lived a real human being, who was probably just run down with responsibility. From that day on, I began to respect and like her and enjoyed my time there.

We had only one day off per month, a fixed day on the ward, and if one got a shift before the day off, it was lost. However, it could work the other way; one might move to another ward and fall into a day off. On Sunday one either had a long morning till 1pm or a half-day from 2pm. There was no time to have any life outside the hospital for the first year, as one worked till 8.30pm then had our evening meal, prayers and a hymn. We did however get a few hours off in the afternoon. Only in our second year did we get an evening off, probably a good thing, as we had no money to spend on entertainment. At the end of February we got our first pay, thirty shillings, and most of us sojourned to the nearby Kenya Cafe for afternoon tea and a bun. How good it felt, just to sit down and have a cup of tea away from the hospital atmosphere.

All too soon, we were on our first night duty and in my case it lasted six months. If one did anything wrong on night duty, like making a cup of tea, one was immediately taken off as a punishment. I never could see where the punishment came in, but although I made tea like everybody else, I was never caught. I had a tiny little suitcase in which I carried two cups of tea from the ward kitchen. My hearing was very acute and just one little sound from the doors in the corridor and I knew that sister was on a round of the hospital, down went the lid on the case and I advanced to meet her, no problem. It was a very long night. We slept out at a place called Woodburn and got a bus into the Royal to start duty at 8.30pm and we did not finish until 8.30am. After we had supper or breakfast — whatever one might call it — we might very well have a lecture before catching the bus for Woodburn. When one did arrive there, one just fell into bed and slept immediately. The one good thing about night duty was one got three nights off at a time. Most nurses went home for this. I was too far away, but often got invited to someone else's home.

My holidays came round in November. Most girls had one week and then three weeks. Mine was all at once; I could not afford to travel twice, it was too expensive. What a joy it was to be at home again and to meet up with all my old school friends, but all too soon it was over and it was back to work.

Early in our second year we sat our preliminary examiniation. It was not too difficult and, if I remember correctly, we all passed. After we finished our exam, the sister in charge gave us a cup of tea and two rich tea biscuits, a rare treat in wartime.

Life was beginning to be a bit more cheerful now, as we did get an evening off now and again. One big event for me was a visit to the theatre. The theatre management sometimes gave complimentary tickets to the staff nurses. One such nurse had an extra ticket and invited me to come with her. I was just spellbound. What lovely singing, particularly the song, "Those two eyes of blue still smiling through at me".

Early in my training one of my sisters married a farmer. I was a bridesmaid and had a lovely midnight blue velvet dress. Unfortunately it took practically all my clothing coupons, so I could not buy anything new, even if I could have afforded to.

Life in the wards was very busy. There seemed barely enough time to get the work done, and sometimes one got blamed for things that could not be avoided. One such time, when we were doing a linen inventory, I came out from the sluice with a damp pillowcase, to add to the count and deposit in the laundry basket. Sister barked at me, "Why was that not in the laundry before?" To her absolute horror, I barked back, "Because the patient had not been sick on it before."

Life was never dull, as we moved from ward to ward — medical, surgical, gynaecology, ear, nose and throat, venereal disease and theatre. As we moved up in our training, we got more responsibility. One such thing was to look after tonsillectomy day cases. Ten were usually done in a morning and as we had only a side room available with five beds, we had to top and tail them, two to a bed. They came back from theatre so quickly that it was a job to get a sick basin to each head in turn.

The ward I liked least was the venereal diseases ward. We had about half a dozen babies in cots with congenital syphilis. Naturally, they were bottle fed and it took such a long time, as we were not allowed to leave them with a bottle in their mouths. I also had a fourteen year old girl with congenital syphilis. She had the most lovely blonde hair, alive with vermin. Sister instructed me to shave her head with the cut throat razor. It was one of the worst jobs I had to do. She cried at the loss of her beautiful hair, and although I explained to her that it would grow again, and no longer be itchy, I could not comfort her. She died a few weeks after I left the ward and I felt quite vexed.

One time a Shetland nurse and myself had nights off at the same time. She asked me to come with her to spend the time in a friend's house. We were supposed to stay only with relatives, so she put down that we were staying with a grand-uncle who lived in Edinburgh. Unfortunately a sister of mine, now fully trained and nursing at Peel Hospital, had got a day off as well. She travelled to Edinburgh to see me and the sister in charge gave her the elderley man's address. She sent a telegram to him to say she was coming, and then she arrived. He said he did not know what to make of it, first a telegram and then a lovely young lady arrived.

Much later, when I was on my third night duty, this same sister came to Edinburgh to spend some time with me when I had my nights off. She meant to stay at a bed and breakfast, but we were too late starting to look for this, and every place seemed full up. In the end we decided she would just have to spend the night with me. We had no bother getting her in past the sister on the door. I instructed her just to keep her head down and say goodnight sister. I had previously bought some buns from the visitors stall, a very forbidden practice for nurses. What we did was walk quickly past the stall, laying down a sixpence or a shilling as we passed. When we walked past again, there would be a bag at the end of the stall with our buns. This was quickly concealed under our red capes, making sure no-one in authority was looking. There was no problem making tea in our rooms. Every fortnight we were issued with our tea and sugar ration, and milk was put out at a geyser for us. My sister and I had a nice cup of tea and then decided to go to bed. Unfortunately a mouse had smelt the buns and I had to get up and put the remainder of the buns into a drawer. That quietened the mouse. Next a repeater alarm clock went off in the room next door. I got up to see if I could stop it, but although the door was open, the clock was in a locked drawer and we just had to listen till it decided to stop of its own accord. I thought that would be the end of the diversions, but no such thing. The air raid sirens went off. Now what? We could not go down to the shelter, so just had to remain. I knew night sister would come knocking at the door so I got up and took the light bulb out of its socket, in case she opened the door with her pass key. She knocked but did not open, thinking I had sojourned to the shelter. Shortly after this we heard gun fire. I think the Germans were trying after ships at Leith. About 4.30am the all clear went and we managed a little sleep before my sister had to get up to catch her train.

We, as nurses, took turns at firewatching, but I never happened to be on this duty when the sirens went off. At this stage in the war things were very bad in London and we were informed we had to try and find beds for men whose Old People's Home had been bombed. The medical ward I was in received six, all with their names pinned onto their clothes. We had to put camp beds down the centre of the ward to get them in. They seemed to have no possessions, except for one gentleman who had four pennies in a watch case. He counted them every morning and every evening. One of these men, who was a bit more "with it" than the others, seemed a bit upset one day. I enquired if anything was wrong and he replied, "The barber has been and I did not have two pennies for a shave." Well I replied, "I do not have two pennies on me either but I have the use of a razor and when I can spare a minute I will do it." The grateful thanks I got when the job was done made me feel quite humble.

As charge nurse I was informed by the doctor that he wished me to try and get a few of these elderly gentlemen mobile. The first one I tried to get out of bed was a retired lawyer, with a double-barrelled name, definitely out of the top drawer. Well, I put a chair at the side of his bed, explained what I was going to do, laid down the bedclothes, put my arm round his back for support and tried to get him to move. Goodness! He let out a terrible roar, "Young lady, you are not taking advantage of me." The ward was in titters. I did finally get him out of bed, but as I passed down the ward, one patient looked up and said, "Nurse, you can take advantage of me anytime you like." One sensible but very crippled patient was so overcome when I got him out of bed he cried. We were able to do a lot for him, just with regular movement and hot wax. No artificial hips or knees in those days.

How we have moved on in medicine, nursing and occupational therapy. During my nursing in the war years I could never have visualised heart, lung, liver or kidney transplants. We did, just after the war, get to grips with TB, but now it appears to have developed a resistance to drugs, and reared its ugly head again.

My next ward move was to Surgical Out Patients Department. There one never knew what to expect, we were always busy. Accidents came there first, before being transferred to a ward, if they required to stay in. Head injuries always required shaving, before being stitched and dressed, and I became quite an expert with a cut throat razor. Then one day a young gentleman (on thrice weekly dressings) handed me a small packet. I said, "What is this? We are not allowed to accept gifts." He replied, "I think it would be quite in order to keep this one." When I opened it, I found a lovely new safety razor and extra blades. What a help it was. Next week he brought me two lovely peaches in a little box saying, "This is so little, it can't qualify as a gift, you can easily keep it." Just fancy, peaches in wartime when one could hardly get a banana.

Naturally we had to take patient's names and addresses, age and occupation. One young lady refused to tell me what she did and I informed the duty doctor I had to leave that space blank. He gave me a sly smile and quickly said, "Nurse, she is a member of the oldest profession."

Time marched on and in our third year we had a little more money. Those of us from the Isles had to save for our fare home. I opened a post office account and put ten shillings in every time I got my pay. One time, practically at the end of the month, a Shetland nurse and I decided to take a walk on our afternoon off, as we could hardly afford the tram fare. We were walking down the Bridges, past a registry office, when a young man in uniform stepped out and asked us to witness his wedding. He and his bride-to-be looked very happy and we were glad to be of help. When the ceremony was over, he presented us with half-a-crown each. We protested, but he insisted we take it and drink their health, even if it was only a cup of tea. They had to catch a train or we could have celebrated together.

Looking back, it seemed that although we worked very hard, we were well treated. It was drummed into us that we were young ladies and must behave as such. We must never demean ourselves to carry out the pig's pail and must never be seen with a brush in our hands. At 12 midday, if one was an inside probationer, one mopped the ward floor, but definitely the sweeping brush was for the maid.

I always felt that Edinburgh Royal Infirmary was well managed, and that the good management came from the top and down.

On night duty, although only two nurses to a ward, one charge nurse and one early in training, one could always get help if pushed. This came about by the fact that every time we started night duty, we had about two weeks without a ward. During this time we sat in a duty room with our nursing books until sister would come through and say help was required in ward so and so, as there had been a death or a big accident. If a few of us were still there in the morning at 4.30am, we were sent to the busiest wards to help fill hot water bottles and make beds.

Once per week each ward had a waiting night. This was taking cases, brought in through the night. Once on my second night duty a very bad accident case was brought in. She went through to theatre and came back down looking terrible. The surgeon actually came back to the ward with her and said, "Do what you can to make her comfortable. I have no hope. Her spleen was ruptured, we had to remove a kidney and she is pregnant." I immediately rang the duty room and sister said, "Have you had a death?" I replied, "No, I wish to try and prevent one if I can. I have a patient back from theatre. She is sweating terribly, I wish to sponge her with the least possible movement and it will take two and my assistant is busy looking after other cases." Sister said, "I will send help immediately, and you may keep her till morning." We did not have penicillin in those days and we relied a lot on real nursing care, which we always tried to give. We did manage to make this patient more comfortable and she was still there when I went off in the morning, and still there when I came on again at night. After about a week she lost her baby and was removed to

Simpsons Maternity Unit. I don't know what happened to her after that.

We sat our finals about the end of our third year or the beginning of the fourth, and then there was a wait of a week or two before we knew if we had passed. Then one day quite a few of us were called to the Lady Superintendent's Office. We wondered what we had done, but it was to inform us that we were being called-up, to go into the forces. Those of us who had passed our finals early in our fourth year would be released a few months early from our training. It was government regulations, she could do nothing about it, so would we decide which of the forces we would like to join. In a way we were glad, as we were going out into the world as trained nurses and would receive a little more pay. I decided to try the Civil Nursing Reserve and went down to their headquarters in Edinburgh where all they asked was my name and where I would like to be posted. They said they were pleased I had decided to join them. For my posting I requested Orkney, explaining my parents were elderly and I wished to visit them as much as possible. However, when my posting came, it was not for Orkney, it was for Tingwall Military Hospital in Shetland. It had been converted from a manse, with a few nissen huts, for wards and staff.

Getting there in the first place was a big problem as the roads were almost shut with snow. Every motorist seemed to carry a shovel and, more often than not, required to use it. The work in the hospital was not difficult, most patients were convalescent and just required ordinary nursing. Unlike any hospital I had worked in, there seemed time to sit down, so I decided to do a bit of knitting. I was, of course, on night duty on my own. No sooner had I settled down, than a big rat came out from under the sink. I screamed, all to no avail, my patients were asleep. Next day an ambulance driver, whom I had been at school with, brought me a rat trap. That evening no sooner had I settled down, than a rat came out and immediately went into the trap. Then all hell seemed to be let loose. The dying rat screamed and rats seemed to come from everywhere. I ran to the male ward and woke up a mobile patient to help. I think he came from Burra. He soon killed the rat, but I informed matron the next day that I was not prepared to stay unless something was done about the rats. Something was done, a big rat cage was borrowed from Nelson's farm and when I came on next evening it held about nine rats. The rats had all been killed or deserted after that, and life went back to normal.

After a few weeks Mr Lamont, Surgeon Consultant, came out to do his rounds, recognised me from my probationer days and said a mistake had been made, I was supposed to be working in the Gilbert Bain. One of the sisters there had been ill and required light duties, she would change with me. Neither of us wanted to change, but after a few weeks we would not have wanted to change back again. Although we were extremely busy, life was interesting in Lerwick. Every Saturday night there was a hop in the Town Hall and invitations poured in to R.A.F. functions etc. On my days or nights off, I was able to go home to my parents and also meet my friends and school chums.

There began for me a bit of young life, which I had missed out on before. In my training days in Edinburgh the only dance I had been to was the probationers' fancy dress one, all female except one gentleman who played the piano. There was no shortage of partners but I could not do modern dancing, so quite often refused. At one of the hops I met my future husband. A gentleman, whom I took to be a Norwegian because he had on a Fair Isle jumper in the Norwegian style, asked me to dance. We danced a bit and then suddenly he spoke perfect English, and asked where I came from. I replied, "A small place you would not know", never thinking he was a Shetlander. He said, "Try me." "Well, actually I belong to Bigton, but you will never have heard of it." He then asked if I knew Jesse Goudie from Bigton, who had been in his class at the Institute. I was able to reply that I knew him very well, as he happened to be my brother. After that we seemed to talk quite a bit but I refused when he offered to see me home, and explained that I was going to run home and could not be bothered with escorts. In actual fact, I was on night duty so had to leave before the hop ended to be on duty for 11pm. I said goodnight to the girls I was with, got my coat and started to run along the Hillhead. Next I heard footsteps behind me and there was this gentleman again. He soon caught up with me and said there was no need to be frightened of him. I started to laugh and said, "The only thing I am frightened of is a mouse or a rat, but I must not be late for my work." I explained that I had it worked to a fine art, so many minutes along the Hillhead, a few down the brae, a few to don the uniform, one minute for any emergency, and then to the Gilbert Bain Hospital. There was no time even to say goodnight and I never expected to see him again. However, this was my last night on night duty at this time, so on Sunday morning I walked up to Ganson's to get the mail bus home for my nights off. As I approached from one side, here was this same gentleman approaching from the other side, to get the mail bus for the west side. He wanted to know when I was coming back to Lerwick and I explained that I started on day duty on the Wednesday, but had to come back on Tuesday as there was no transport early enough on Wednesday. On Tuesday evening, about 7pm, he came to the house where I stayed and said, "Get your coat and prepare to run again, there is an R.A.F. lorry at the Town Hall waiting for us. We are going to one of their dances out the North Road." Sure enough there was the lorry and a young R.A.F. man jumped down and helped me on board and we arrived at the dance. That was the start of things and we married in 1947.

# CHAPTER 3
# Preamble to "Lang Trachled"

AS "Lang Trachled" is about life in a shop and post office, I now write a short description of a Shetland country shop, just after World War Two.

In January 1947, when I joined my in-laws helping in such a shop at Bridge of Walls, it was very much like a store in a wild west film.

It was also rather like Ronnie Barker's shop in "Open All Hours". The exception being that the till was just a drawer in a table, and fortunately never jammed!

Most things came in bulk; tea in a big tea chest — being weighed up in brown paper bags in the quantity the customer required. If one returned the empty chest one shilling and six old pence was deducted from one's next account. This rarely happened, as the chests were much sought after. They were just the correct size to set a clocking (broody) hen. Most customers got the tea chest free.

Sugar came in coarse hessian bags, which when empty were used to bag peat or blue clods — blue clods being small pieces of best quality peat.

Butter came in a small cask, called a firkin. I believe it held 56lbs of butter, which also had to be weighed out in half or whole pounds.

Bacon was bought in a roll or a streaky side. A very sharp knife was required to cut some. In fact it was sharpened so often it just sort of wore away and a new one had to be acquired.

Many other things had to be weighed up; lentils, barley, rice, flour, oatmeal and baking soda. Rough salt came in a barrel, which when empty could be used to soak dirty clothes — no washing machines in those days, in fact not even electricity!

Brown and black tobacco also came in bulk, and was referred to as buggy roll. On the shop keeper's side of the counter there was a nick in the wood which determined the length to be cut for one ounce. It was usually correct, seldom under or overweight. Over the shop door hung a big notice proudly declaring that Thomas Hobbin, Draper and General Merchant, was licensed to sell tobacco.

Paraffin was in a shed outside, as were the measuring jugs for dispensing same. People brought their own containers, sometimes only a bottle. One such customer set his bottle down on the hard shop floor, where it fell over and broke. "Oh", said the customer, "that silly old bottle has leaned over and broken its neck."

Bridge of Walls shop was a long way from the chemists in Lerwick, so lots of cough and stomach mixtures were kept. Castor oil, Epsom Salts, Beechams Pills and Syrup of Figs for constipation, chemical food to increase the appetite after any debilitating illness, and some brand name of liver pills that were supposed to wake up one's liver bile. Venus cough mixture and chloridine sweets were on sale for sore throats. Pink and white lint for infected cuts, although available, was seldom bought. I understood this from my own childhood because any dirty cut was treated with a loaf or carbolic soap poultice. Root ginger was also sold for general nausea, or those wishing to take a sea journey, and aspirin for a sore head. Rennies were used to relieve stomach distress after any extra indulgence. Rock sulphur was much in demand to whiten shawls and also for cramp. Apparently if one got cramp, one held a stick of rock sulphur in one's hand for a few minutes and got relief from the cramp. In my young day a mixture of brimstone and whisky was given to bring out the rash in the case of measles.

Many Shetland people had their own remedies. One such person told me he kept a raw potato in his jacket pocket to ward off rheumatism. He said it really did work. If he changed his jacket, and forgot to transport the potato to the new jacket, back came the rheumatism. Razor blades and Brylcreem were on sale for gentlemen. Toothpaste and shampoo for all, but except for hair and hat pins and kirby grips, little was available for the comfort and enhancement of ladies.

I suggested sanitary towels should be on sale, but this was frowned on by my sisters-in-law. They explained that no lady would ask a gentleman server in a country shop for sanitary towels. It would be indelicate. Changed days now I think.

The needs of the scribe were also catered for. Writing paper, envelopes, jotters, brown wrapping paper, string, pens, pencils, plain postcards and picture postcards were all on offer at what would now appear very cheap prices.

The drapery side seemed rather scanty, but then coupons were still required for clothes. As I remember each person had a yearly allowance of sixty-six clothing coupons. A woman's wool dress required eleven. Two ounces of knitting wool required one coupon, as did two handkerchiefs. One had to be in real need of a garment before deciding to part with the coupons. Nevertheless, there was plenty of underwear, long johns, vests, socks and fleecy knickers. Dungaree trousers and jackets hung from the rafters. Night dresses, the thick wincyette type, and night shirts were kept in a drawer. Another drawer held shirts and collars for gentlemen. There was quite a supply of rubber boots, slippers and shoes plus leather and sprigs for mending.

Household goods were also stocked — brushes, shovels, long fire tongs, towels and dish towels. Before the war there had been a good supply of cups, saucers and plates, but alas they had all been sold. There did still remain a few iron frying pans.

Most people repaired their own windows, doors and gates, so nails, staples, putty and glass were often in demand. Rope and coir were bought to secure corn and hay stacks.

On the agricultural side, Indian meal and the seeds of corn were sold for hens. Sheep did not seem to require much in the way of bought feeding, but sheep dip was sold to keep them free from ticks, and a drench to prevent liver fluke.

The most usual colours of paint and distemper were kept, plus brushes to apply. Tar was kept for waterproofing of roofs.

Usually one Shetland spade, one hay fork and one tuskar (a spade for cutting peat) were on display. When one sold, another was ordered from Lerwick.

Small hand lamps, tilley lamps and spares brought in quite a bit of money before electricity arrived.

As no calculators were available, one added up in one's head or used a bit of scrap paper.

At one time when my husband and I were the joint owners of the shop, we did a barter system. A representative arrived requesting orders for linoleum and rugs. My husband explained that he could not place an order. He had a big overdraft, having bought his sister's share in the shop. In those days there was no Shetland Islands Council giving grants or loans at small interest. One just floated or sank, by one's own effort. "If you happen to be back here later on in the year, when I have sold my summer's clip, I might place a small order", explained my husband. The representative was most interested, "What was the summer's clip?" he asked. When it was explained to him that it was mostly Shetland wool, some Blackface, plus a little first cross Cheviot wool, his interest increased. He offered 15 per cent more that we could get in Lerwick for the wool, if we would take linoleum and rugs in exchange. Also he would pay the carriage on the wool and linoleum. A deal was struck and it was an absolute Godsend, as most of the floor coverings at Bridge of Walls House just had to be replaced.

We quickly sold the rugs and linoleum and could have done business another year, but had to decline. The quality of floor covering at that time was so good it lasted for years and as our customers were the same from year to year, none was required.

At this point I will digress for a moment to relate a handed down story. A crofter went to Lerwick to sell his wool and when his children saw him arrive home they ran to the house in excitement to see what dad had brought home for them. Unfortunately dad had been celebrating and was asleep in the resting chair. Their mother looked at them and said, "There he lies with the bulk of the summer's clip inside him."

As regards being "Open All Hours", we could mirror or exceed Ronnie Barker's opening hours. In the country one came any old time, especially if busy during the day with croft work. Very different from the supermarkets of today, no self-service but lots of interesting gossip, definitely a social place.

However it was very hard work and I for one have no wish to do it over again, even though some might call it "the good old days".

# CHAPTER 4
# Lang Trachled

ON the 29th January, 1947, I married Norman Hobbin. We had some trouble getting a cake made. One of my ex-patients had promised to bake my wedding cake if I could get him the sugar, fruit and eggs required. He worked in a bakery owned by his brother. Well I managed to get all the ingredients with help from my own family and aunts in Quarff. My husband-to-be went along to the bakery and was met by the owner. No luck, he declined to bake the cake, in spite of being informed that his brother had promised. He said they just did not bake cakes for anyone, and his brother was not him. Next port of call was Black's Bakery, who supplied the bread for the family shop at Bridge of Walls. On enquiring there, the answer was the same, no wedding cakes baked. However, as a matter of interest, the owner asked who was getting married. "Just myself", answered my husband-to-be. "Good gracious me", said the baker, "in that case an exception will have to be made. You and your parents before you have been customers of this shop for many years; I will bake the cake." Since he was getting all the ingredients, an extra special one would be made. It was too, and the top tier was as good as ever on the 2nd December that same year, when my elder daughter was born.

The actual reception went off quite well in the Bigton School. It was just sandwiches and tea, with a little spirits thrown in, as food was still rationed. Curly Jamieson's band from Sandness supplied the music and a good night of dancing seemed to be had by all. We actually had a second night. The people in the district laid together and provided the food and music for a party a few nights later.

On the way home from the reception, there was a little flurry of snow, a forerunner of things to come. We had about two weeks at Bigton, just managing to get to Bridge of Walls House, my new home, before the roads blocked with snow.

I started my married life in a house with two other couples. My two sisters-in-law were both in residence with their husbands. Strangely enough we got on very well. We cooked, washed dishes together and sat around the fire in the evening together. We often had a musical evening, my husband or brother-in-law, on the fiddle, and a sister-in-law on the piano.

As the roads were all blocked no food could come from Lerwick. Fortunately we discovered that Mr John Hay, the lone fisherman, was fishing out of Walls. My husband and a brother-in-law took a small sledge and walked to Walls for fish. We had fried fish, boiled fish, steamed fish and baked fish, till all the fish were finished. Then the fish heads, which had been soaked in salt, were washed, stuffed with oatmeal, baked in the oven and eaten for dinner.

This home did not have a barrel of salt meat like most Shetland houses at that time. When I lamented about this, my husband said that there was a big bone in a basin outside in a meat safe. He duly brought it in and cut it in two with a hacksaw. I went to the back kitchen for a cooking pot and when I came back, "Gandie" the cat, was on the floor with half of the bone. I was very cross on the cat, but made a good pot of soup on the other half and everybody enjoyed it.

After we had experienced the blocked roads for quite a number of weeks, it was a big treat when a boat came into the pier with mail and rations. One brother-in-law went back to Lerwick with the boat, so that he could carry on his work as a car engineer. He also repaired wireless sets. One such set had a letter with it from the owner, saying the set was not working properly, "Sam as she was shocket!" I think Shetland words explain things so well.

Soon my other brother-in-law went back to sailing and his wife got a post as a teacher in Sandwick, returning to the Bridge of Walls at the weekends.

Very soon my life was taken over by preparations for the birth of my first baby. This took place, as I said earlier, on 2nd December 1947.

The summer of 1947 was bright, warm and sunny, after our snowy winter. Otherwise things were not so good. In September my husband became a guest of His Majesty over a motoring offence. There were no Social Services in those days, but my own family rallied round, as did my ex-colleagues. One even offered me to come and stay with her while my husband was away. Fortunately my husband saw I was not short of money so I did not require to seek charity. He returned home on 30th November so was able to escort me to the hospital for the birth of our first baby. The next year was also rather sad as both my father and mother died; father in June and mother in December. However there was not much time for grieving. I had to prepare for the birth of our second child, who was born on the 18th January, 1949. Life for a time after that was just a case of looking after the children and sometimes serving in the family shop.

A year or two after this, both sisters-in-law said that they would be moving out within a few months. One's husband had got a shore job in Newcastle and when he found somewhere suitable to live she would join him there. The other's husband had been lucky enough to get a job as an engineer to the Herring Board in Lerwick and a flat went with the job.

They offered to sell us any furniture they did not wish to take with them. My husband had inherited the tenancy of the croft and house at Bridge of Walls, but most of the furniture had been left to his sisters. One thing I personally bought was a big telescopic table which had three leaves and when extended could seat fourteen people. A big sideboard went with it and both were of immense value to me later on when I catered for summer visitors. We bought quite a lot of other items but one we did not buy caused a bit of fun. This was an old fashioned kitchen cupboard which the late parrot had chewed. Norman, my husband, offered thirty shillings for it, but it's owner said two pounds or nothing. I was then asked for my opinion so I said, "One pound or nothing, I don't like that old cupboard." "Yes", said my husband, "but it can hold a lot of plates." My reply, "So can two orange boxes nailed together and camouflaged with a bit of curtain. That will do until you get some wood and make us a proper cupboard. It is a pity we are so landlocked here, otherwise we might have got some driftwood." The cupboard then went to the sales in Lerwick where it fetched fifteen shillings and cost one shilling and sixpence for transport! As it happened, Norman did not need to make a cupboard as I was given the home-made dresser from my old home.

Once my sisters-in-law and their husbands had moved out the business of trying to get Bridge of Walls House back into some sort of order began. Nothing had been done since before the war and only one oven in the big kitchen range could heat up enough to bake in. Apart from that it required a kishie full of peat just to get it started in the morning.

I explained to my husband we would have to do all the painting and papering ourselves to save money. He had had to buy over two thirds of the shop stock, plus two thirds of the sheep on the croft when his sisters left, so money would be tight. Unfortunately much more than paint and wallpaper were required. One night we were lying in bed, almost asleep, when we heard a loud bang. My husband went downstairs to see what had happened but all was in order, and no burglars. Next night at precisely the same time, twelve midnight, again a bang and then a rumble. Next morning I got up, came to the top of the stairs and could not believe my eyes. The ceiling above the stairs had come down. It was lath and plaster and a whole lot of stones and rubble had come down as well. What a blessing no-one had been coming down the stairs at the time. Norman had to get the barrow and clear it all away. Now we had extra work, but Norman said no way could he put up plasterboard. "In that case," I said, "we will call on extended family." "Extended family Agnes, as I understand it we have a little boy not quite two and a little girl not quite three", my husband replied. "No, my own extended family is what I mean, I will get a message to my brother in Scalloway and he will instruct you."

Well, after consultation on the phone with my brother, we ordered and got the plasterboard. Next my brother arrived and between him and Norman the job was done. Afterwards they inspected other ceilings in the house and said the dining-room ceiling must come down, before it fell down. This ceiling Norman did himself with a little help from me carrying in the plasterboard.

One thing we did require help with was the installation of a new range. We were lucky to get a plumber locally willing to do the job. Also lucky to get a shop discount on the range and in no time at all it was working in our old front kitchen.

Our biggest job was the laying of a new floor in our front kitchen. We discovered a wooden one would be quite expensive so decided on a cement one. After all, we had a little boat to fetch shingle and a car to fetch sand and cement. Trowels and a spirit level were already about the place. As it happened we were lucky as my brother Jesse, his wife and family, paid a social visit on the Sunday. I think Jesse wanted to cast his eye on the dining-room ceiling, and when he did, declared the job well done. He spent some time instructing Norman how to go about laying a floor and said, "Lay the last covering of cement on a Saturday and I will come and skim it off for you." Preparation for the laying of our new floor was quite good fun. I would lever off the old boards with a crowbar and Norman would carry them out. He wondered if we would find anything under the floor. I then had the bright idea of dropping down two half-crowns and when he lifted the next board they rattled out. My fun was short lived, as he looked at the dates and said, "Good try Agnes, but this floor was laid in the 18th century." We did find lots of things which must have dropped down between the boards before linoleum was laid. The collection included beads, small toys, half a pencil, a quantity of used razor blades and some small bones. The razor blades disintegrated when we tried to lift them. The bones we concluded had been from two generations back when people were in the habit of keeping and feeding their dogs inside. We finally got everything cleared, the new floor laid, and Jesse came from Scalloway to give the finishing touches.

Then began the big job of hanging wallpaper and painting doors, skirting boards etc. Norman did all the painting, but I helped by pasting the wallpaper. We got up early and worked late as we had the post office and shop to run, the animals, ourselves and the children to feed.

Now we required quite a lot of furniture — beds, wardrobes, dressing tables, ordinary tables and chairs. I journeyed to the auction rooms in Lerwick and it proved quite fruitful. The North of Scotland transport brought home my purchases a few days later. One thing I was very pleased about was a big wardrobe I had acquired for a few pounds, the bidding for it not being strong. My husband took one look and said, "No wonder you got it cheap; it is too big to go upstairs." This was a big

1. Geosetter croft in the background, viewed from Bigton farm, c.1901.

2. The author's father and half-sister cutting peat in the hill above Geosetter, c.1912.

3. Bigton church.

4. The author's brother and sister-in-law carting hay, c.1935.

5. The author with brothers and sisters during her first year at school. From left, back row: Jesse and Robert. Front row: Betty, Annie Grace and myself, c.1927.

6. The author at Geosetter burn, 1999.

7. The Gilbert Bain Hospital (1902 - 1984).

8. With colleagues at the Gilbert Bain Hospital, c.1939.

9. Nurse Training at Edinburgh Royal Infirmary, c.1940.

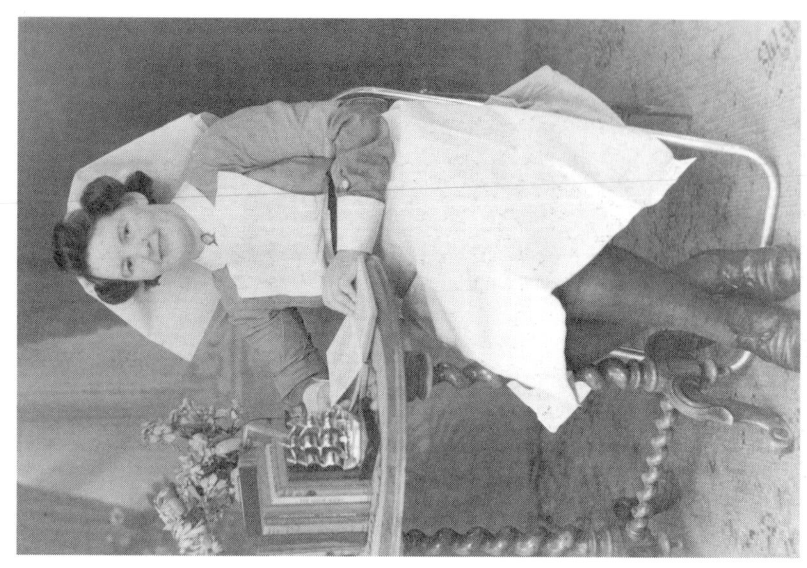

11. The author wearing indoor uniform, Civil Nursing Reserve, c.1945.

10. The author wearing outdoor uniform, Civil Nursing Reserve, c.1945.

12 and 13. Bridge of Walls house.

14. Sailing boat at Bridge of Walls.

15. Thomas Hobbin, the author's father-in-law, (centre) with two of his first cars, Norman Hobbin (left, in car) and George Hobbin (right).

16. Boats used for summer visitors, tied up for the winter.

17. The author's brother-in-law, Dr Gordon Hobbin, with his catch after a days trout fishing.

18. Norman Hobbin in his lobster boat.

19. A dinghy built by the author's brother, Jesse Goudie, at Bridge of Walls pier.

20. Agnes Hobbin at her graduation, Dundee College of Technology, 1976.

21. Retirement party, 1982. Agnes Hobbin front row, third from left.

blow, until we discovered that it could come apart. Even it two pieces it was still a struggle to get it upstairs. Next I had a quick trip to Aberdeen to buy what I had not been able to procure in Shetland. Furniture was still rationed but we had enough coupons for a living-room suite; everything else had to be second-hand.

We had decided that once we got the house looking reasonable we would advertise for summer visitors. At last we were ready, even all the beds made up, and visitors began to arrive. They were nearly all anglers except for their wives and a few birdwatchers.

We were very busy. We still did not have electricity and every bedroom had to have a lamp. I was able to employ two maids and they worked very hard indeed. Very soon people arrived asking for meals and teas, especially on a Sunday, and we did them as well. I baked scones, rock buns, little buns and currant squares, ready for Sunday afternoon. More often than not, I had to get the baking bowl out again and bake pancakes to keep the tables going.

That summer my husband was at home and did most of the shop and post office work, and filleted the fish the anglers brought in. They nearly all had brown trout before their bacon and egg. All sea trout were boiled and eaten with salad. In the garden we had our own lettuce, carrots, turnip, beetroot, peas, mint, potatoes, parsley, rhubarb and cabbage. There was very little time to weed the garden. Sometimes an angler's grass widow would offer to help when she was bored. I always said, "Feel free, a little bit of help is better than a great deal of pity."

We had our own fresh eggs. I had acquired a chicken brooder from my old home and early on had bought fifty day-old chicks. They were now in lay, along with a few old hens. Fresh milk was in abundance as we were able to hire a cow from Mr Duncan Houston for the summer. This was a sort of barter system, no money changed hands, we fed a young animal at the same time. The cow we got that first year was a very pleasant cow, appropriately called Daisy. The children loved her, she was so peaceable, they often used to stroke her.

The next summer my husband got a job on a water scheme in the Highlands, so I had more responsibility. Fortunately I was able to employ a mature lady, willing to milk the cow and carry in peat. We had a Truburn stove going in one end of the house and a Rayburn stove in the other. The cow we had that summer was a big Ayrshire. I was wary of her but was obliged to milk her on my senior maid's day off. I always got one of the other girls to come with me for moral support. Between us we would tie her to a Hydro pole while I did the milking.

My duties were many and varied. One time we had a doctor staying who got a foreign body in his eye. He enquired of one of the girls where the nearest doctor lived, but she said, "Just go through to Mrs Hobbin, she is a trained nurse." It proved to be no problem, with the aid of a matchstick I was able to invert the eyelid and remove the object. This same doctor when he paid his bill added £5 to it. I said, "You have made a mistake." He answered, "Not so, at Harley Street it would have cost much more that £5 to get treatment for my eye." He then went on to say he had never met anyone who could do so many different things. "One day you not only changed the washer on the bathroom tap, but made the new washer out of an old rubber boot. Then a few days later, I was in the post office, when one of my fellow anglers asked where the nearest mechanic could be found. His self-drive would not start. You calmly carried on serving your customer and said, 'Fords don't like this misty weather. What you need is a stick and a bit of cloth. Dip this in the petrol tank, take out the distributor cap, rub it with the cloth, replace it, wait seven to ten minutes and try again.' This was done and the car started no bother. Then there was the question of anglers wanting to send trout home. You instructed them to roll them first in cabbage leaves from the garden, to keep them cool. Next paper and string. Lastly bring them to the post office, where you would, for a few extra pence, send them airmail special delivery. A postcard arrived a few days later from one recipient saying the trout had arrived in good condition. I know you do the cooking. I have seen you filleting the fish, milking the cow and on Saturday night in the post office doing the books." I explained this was just what was expected of Shetland women. Most of the Shetland men were sailors, being away for long periods. Any Shetland wife just knew she had to keep the house and croft going till her husband returned from the sea.

A very busy day in my life was when there was a cattle roup on. This was actually held on our croft. Everybody was buying sweets and cigarettes and quite often banking money, if they were selling. In those days most crofters used the Post Office Bank. It was much handier than going to Lerwick. I did lunch for the auctioneer and his helpers. Teas were served for the buyers and sellers. One elderly crofter said, "It is grand to have a cup of tea but before your time we used to have a better drink here. We would gather in after the sale for a drink and a chat. I remember a minister that used to come as well. We said we were surprised to see him having a drink. His reply was, 'Not that which goeth into the mouth defileth a man — but that which cometh out of his mouth defileth a man'."

One roup day, when the post office had five tall young gentlemen talking together outside the counter, a very old lady came in. She was dressed in old-fashioned black clothes, currently worn by her age group at that time. She had a lot of pension money to collect, having kept it the regulation amount of weeks one was allowed. When she opened her purse to put the money in, it was already full, she having sold an animal. One young gentleman leaned forward and asked what she was going to do with all that money. Quick as lightning she said, "I might go on holiday." "Oh, where to?" asked someone. "Perhaps to Edinburgh first and then up to London." Now it was quite obvious she had never been out of Shetland and had no intention of going, so she was asked what she would do in London. Quick again came the answer, "Go to the theatre and music halls and perhaps end up going on the game." No more questions, just some deflated young men.

Christmas was another busy time at the post office. When I was sub-postmistress, postage was not as dear as it is now. Lots of people put legs of mutton, chicken and turkeys through the post. The mail order catalogue was also widely in use for ordering Christmas presents. In my time in the post office, I only once got into trouble with the Head Post Office in Lerwick. I had received new scales for weighing airmail letters. I set them up for use and threw the cardboard container into the wastepaper basket. This happened at my most busy time, when the house was full of summer guests. A day or so later the overseer working in stores rang up asking why the cardboard box had not been returned. I explained I did not know he wanted it back and now I had not got it. "You did get it", said the overseer. "I know I got it, but do not have it now", I replied. He then said, "You must search for it as I have to send it back to headquarters." Twice more he rang about it and said, "Where do you think it could have got to?" This time I thought, well to hang with it, and sang down the phone,

*"Sailing down the voe,*
*Sailing down the voe,*
*Over to bonnie Browland,*
*Where the siggy flowers grow."*

I then said, "I cannot get it back, please don't bother me about it again. If you do, I might get a sore head and have to buy aspirins." My bit of fun was taken in good part. Next time I received stores when I opened them up two strips of aspirin were on the top.

Cooking for the summer visitors was mostly the old-fashioned country type. Home-made soup, home produced lamb, the best butcher's mince and stew and home produced chicken. One day a retired Royal Naval officer brought in a pail of mussels and requested me to cook them. This was something I had never done, so I worried in case I might poison someone. Then I remembered, I had a half-sister living in Walls who taught in the little island of Havra during the First World War, surely she would know how to cook them. No such luck, but she suggested I ring up the Grand Hotel. The cook there had been to a domestic science college in Edinburgh. She did not know either, but rang me back after going through some cookery books. First I had to tap each mussel sharply and the shell should close. If any mussel stayed open it had to be discarded. Then after scrubbing them, boil them with a little bicarbonate of soda. When they opened, they were cooked. Any that did not open had to be thrown away. One could serve them on one half of the shell or on snippets of toast. I decided on the toast, and did so for the two weeks the Naval gentleman was in residence. A few years later when my husband was at home for the summer, he bought a boat with an inboard motor and some lobster creels. We were then able to add lobster and crab to the dinner menu.

One great help we had before we got electricity to the west side was a fridge run on paraffin. It had become redundant from a school that had recently installed electricity. It was advertised in *The Shetland Times* for offers. My husband suggested an offer of £30 but I said, £31.10 shillings, as most people will offer a round sum. Our bid was successful. When it arrived I did not know how we would get it into the house, it was so big. We had to take the jambs off the back kitchen door and seek help from the mail driver to help lift it in. It certainly could hold its fill, had nine settings, so could turn up to freezing if required. When we did get electricity we discovered we could get a unit and turn the fridge over to electricity. Norman was against this, it would be expensive to get an electrician from Lerwick. Much better to buy a new smaller one. I disagreed, explaining to him we were in possession of a big catering fridge. It was very important when catering to keep uncooked and cooked meat completely separate. "In any case, we don't need an electrician, you can do it yourself. You can read and follow instruction and you are not colour blind. I will make sure the electricity is turned off at the main while you are working, so no need to worry." Under protest the job was done and the fridge worked perfectly. Electricity made a great difference to our lives, as anyone will know who has made enough toast for twelve people with a toasting fork and a peat fire. Ironing also became much easier. The old mangle required two people, one to turn the handle and the other to insert the article for ironing.

Most of our summer guests were professional people. One summer we had five ministers at one time. One such minister, who came back for many years, held communion every Sunday, outside if the weather was fine, inside if not. I had the job of seeing the bread made ready. He used to come into the post office every morning to tap the weather glass, then he would say, "This is the day that the Lord has given, let us rejoice and be glad in it." Now when I am really old, I wake up in the morning with sore and creaking bones and I say to myself, another day to endure arthritis. Then I have a flash, rather like W. Wordsworth and his daffodils. He said,

*"For oft, when on my couch I lie,*
*In vacant or in pensive mood,*
*They flash upon that inward eye,*
*Which is the bliss of solitude,*
*And then my heart with pleasure fills,*
*And dances with the daffodils."*

Now what I see in my mind's eye is the Rev Stevens saying, "This is the day that the Lord has given, let us rejoice and be glad in it." Suddenly I feel an uplift of my spirits and feel able to face the fact that old age does not come alone. He still thinks about Shetland, although no longer able for the journey. My son had a card from him at Christmas.

One day when I was going to Lerwick for messages, my husband said, "The Rev Stevens wishes you to bring back a bottle of whisky." I said, "Likely story", turned quickly and there was the minister at my back. He said, "Don't look so surprised, even St Paul took a little wine for his stomach's sake."

We had many interesting people, one of whom came back for many years with his wife and family and had quite a few stories to tell. His favourite one was about a lady who stopped her car on a yellow line in order to visit a toilet. She was booked and the gentleman on the bench fined her £1. When she protested he said, "You know the old story, in for a penny, in for a pound."

Another story was about a concert where the lights would not work. The manager came on the stage and asked if there was an electrician in the audience. No luck. Next he asked if anyone knew a little about lights. No reply. In desperation he said, "Does anybody know anything about anything?" Up then rises a little Chinese man, comes on to the stage and says, "Everybody put your hands up." They did so and the lights came on. The manager, in amazement, said, "How did you do it?" The reply from the little Chinese man was, "Well you know the old story, many hands make light work."

His third story was about a fishmonger, who was asked for a loan of money. He refused saying, "I promised the Bank of Scotland if they would not sell fish, I would not lend money."

Wednesday was usually a very busy day for us as we provided afternoon tea for the passengers on Leask's bus tours. One such day a gentleman from the bus asked to speak to me. He said he had enjoyed his tea, this was a lovely place, could he please book a room. I was unable to accommodate him at that moment, but was able to take him the next week. He was a doctor of medicine who had been, among other things, at the liberation of the Belsen War Camp. I think this may have had an effect on him. He did not look well and was rather too fond of strong drink. His request in the food line was for potatoes with their skins on, boiled in sea water. When boiled, butter melted over them then dusted with oatmeal. Strangely enough when this dish went into the dining-room, all the guests partook and thought it great.

He had a repertoire of stories, his standard one about two politicians who never could agree with each other. One time after a bad argument one said to the other, "I don't care whether you die from the great pox or the hangman's rope, but I hope it will be soon." The other's reply, "It will all depend whether I embrace your mistress or your politics first." In the evening, when he had rather too much to drink, he would come into the kitchen and tell a story. Next morning he would come in again, start to tell the same story, then stop and say to my husband, "I can't tell you that story, there is a lady present", the lady being me.

Most visitors used to borrow our car from time to time to go to the lochs or Lerwick. One day my husband came in from feeding the hens and said, "Who has gone up the road with my car?" I replied, "Actually, Dr F. has gone up the road with our car, don't you remember dear, the magic words said by you about twelve years ago, 'with all my worldly goods I thee endow'?" A bit of a stillborn smile on an angry face, but also the words, "I would not have lent him the car as he will likely get drunk, crash it or get taken up by the police." I replied, "Well I lent him my half of the car and unfortunately your half was attached, so had to go up the road as well. You have only to look at him to know he has a heart condition. I do not want him trailing round the shops on his feet or running for buses. However I don't think he will damage the car, although I do believe he may have a small drink in Lerwick. His tolerance of drink is greater than most, his hand will probably be steadier with a little drink than without. Regarding the police, he is one of the most intelligent people I have met, so may very well outwit them."

I expected him home for his afternoon tea, this did not happen, nor did he arrive for his evening meal. By this time I began to think I had made a mistake and began to worry about him. At 9pm when I had almost given up hope and endured some strong looks from my other half, I heard the front door opening. I ran and said, "Oh it is you Dr F." "Of course it is, who did you think it would be?" While his meal was being warmed up, he came into the kitchen and said to my husband, "Norman, I have brought a present for your wife." Out of his bag came a big round Edam cheese, plus a little bottle of vitamin C tablets. He explained that he thought I was probably not getting enough calcium and vitamin C for my condition. I was pregnant with my third child. He also said, "You know Norman, you have not even provided an easy chair in this kitchen for your wife to sit on, if she ever has time to sit." My husband informed him he had made an unwise choice. I was allergic to cheese and was unlikely to take the vitamin C. I thought a well balanced diet was all that was required. Regarding an easy chair, because of the Clean Food Act, the law declared no such chair is allowed in a kitchen where food is prepared and served. "Oh", said Dr F., "the law is an ass."

Next Norman asked him if he had been aware of any members of the police. "Oh yes", was the answer. "I stopped on a yellow line in order to visit the toilet. When I came out two policemen were at the car. They informed me I could not park on a yellow line, and I answered in French, that I could not comprehend. One then said, 'Let the old b—— go, he is a foreigner, it would cause a lot of trouble to book him.' I left quickly and resisted the temptation to give them a blast of Glaswegian as I knew I was over the limit." "What if they talked French?" asked my husband. "Then into Russian." His late wife had been Russian.

After this disclosure he had his meal then came through to the kitchen with a pair of trousers over his arm. He wondered if Norman had asked him to buy them. He had been all round the guests and no-one wanted them. He then held them up against himself and said, "I remember now, they are for me."

Next morning he came down in his dressing-gown to say he would not be requiring breakfast. In that case, I informed him, I would ask one of the girls to please take him up tea and toast. He was going to refuse, but I told him nobody was allowed to die in my home from dehydration, as long as they could swallow. If they could not, or would not, medical aid would be summoned. "As you know yourself doctor, tea is a gastric sedative and the toast will be well done to clean your stomach."

My husband was of the opinion there was no point in sending up tea and toast. He would only open the window and throw the tea over our sycamore tree. My reply, "In that case the tree will benefit. I remember my mother keeping left over tea, plus the leaves, to put on the plants she was allowed to keep." My father was a very Victorian health conscious man who declared that the leaves of plants were a trap for dust and insects. "Well", said my husband, "I am not a Victorian husband, you can have as many plants as you like. Come to think of it, there are no plants in our home, why is that?" My reply, "Norman, have you never realised that half my genes come from my father's side?" At any rate when the tray came downstairs the tea and toast had all gone. We presumed that Dr F. had benefited and not the tree. He came in July for two weeks and asked to stay on. We finally agreed to keep him till the last day of November. I explained that once December came in we would be starting to get busy in the post office. Then I would need to be having a rest in the afternoon, in order to produce a healthy baby, due in the middle of January. During this time Dr F.'s health improved, drinking became less and less, and I never saw him drunk again. After he left he ordered quite a lot of Shetland knitting, and it was nice to hear from him from time to time. Finally a relative of his wrote to say he had died in foreign parts and one felt one had not only lost an ex-guest but a friend.

That season we had two other guests who stayed until the end of November, a Dutchman and his English wife. She had come to Britain for a big operation and required a few months rest before going back to Holland. She had relatives she could have gone to but felt my house was like a home from home and she would rather stay with me. When she was not resting or reading she spent her time knitting matinee jackets, bonnets and bootees for my expected baby. The Dutchman produced a lot of fireworks and persuaded Norman to help him build a Guy Fawkes for our two children. I produced fancy biscuits and lemonade and a good time was held by all.

Towards the end of the summer season, Norman went to a house clearance sale. He arrived back with a moquette sofa and proceeded to set it under the front kitchen window. I said, "What price the Clean Food Act now?" "Well", he said, "the season is nearly over and I thought it would be handy any time we don't wish to bother with the sitting room fire during the winter." I agreed, but said we would have to cover it up with a sheet in the summer. I then got the Dettol bottle down, a pail with nice warm water, a stiff brush, a clean cloth and that sofa got a good spring clean. All down the sides I went and fished out various objects, coins amounting to over £1, one old-fashioned fountain pen with a broken nib, one biro pen that could still write, two pandrops, three buttons, two nails and a Woodbine cigarette packet with a squashed Woodbine still inside. That unfortunately was not all the sofa produced. Next evening I screamed. I had seen a mouse. Norman said, "It can't be. There are no holes in the skirting board and we have no mice." However to humour me as he knew I was terrified of mice, he set a trap, not expecting to get anything. Next morning a big fat mouse was in the trap. Norman went outside and brought in our cat. The cat sniffed about for a minute then laid her paw on the arm of the sofa. My husband then got a sharp knife, cut a straight line on the fabric, lifted the piece and sure enough there was a nest of little mice, just about ready for the road. What a field day the cat did have! The sofa looked well, the springs were perfect, but I could not abide it after that. Very soon it was relegated to an outhouse; the children played on it and I sat once more on a wooden chair.

The arrival of our third child took place on 25th January, 1959 (Burns night). That was the big event in our lives that year. She was a very good baby, sleeping right through the night till 6am when her feed was due.

Next summer visitors started to arrive and life went on as usual, till late September. Then one retired Army captain failed to return home for his evening meal. A little while after the meal had been served and eaten I began to be worried. After all he was quite elderly so a search party was organised. They found him trudging along in his stocking feet. Both his waders were full of fish, as was his landing net. He had been to Voxterby. He said the fish were taking so well he felt he could not stop. In all he had forty-six trout, mostly brown, but at least six were sea trout. While I was seeing to a meal for him, he had a nice bath and got on a pair of clean socks and shoes. Then when he came downstairs for his meal he said, "Could you please book me in for the same time next year?"

After the summer season my husband left to spend a season at the whaling in South Georgia. Once again I was in charge of the Post Office, shop, sheep, hens, dog and now three children. I felt obliged to keep on one of my summer workers as I could not leave baby Violet alone in the house while I was outside feeding the animals. Life was a little drab but I did have a little interaction with my customers when they drew their pension or bought a few messages.

One very elderly gentleman who came in to buy tobacco could not remember what kind he smoked. He thought it might be something called God's Glory. I informed him I did not stock anything called that but I had something called Harvest Gold. This turned out to be what he wanted but he complained about the price. He said, "I started on the mossy peat and can see myself finishing on the mossy peat."

Another elderly gentleman, who lived along the shore, related to me his experiences when he went to sea first. He said he felt green, and must have looked peculiar with a home knitted moorit gravit wappit round his neck. Also his bit of best suit was crumpled after the boat and train. Somebody directed him to a ship that might require a sailor and he asked to see the captain. The captain's first words were, "Where do you come from?" and when he said, "Shetland sir", the captain replied, "Is that not where they eat the missionaries?" "I then looked him straight in the eye and said, 'Do you know sir, I think they have given that up now.' I got the job and only then did I realise the captain himself was a Shetlander." This elderly gentleman also told me about how he and his shipmates were taken to court for smuggling. He just fixed his eye on a knot in the ceiling and when addressed answered, "Yes sir, no sir". Then the fellow on the bench said, "Let that one go, he is not quite right." I quite enjoyed his stories about his sailing life.

Only once did he come to my shop a little short of humour. An old moorit ewe belonging to me had got into his yard and eaten some cabbages. "Well", I said, "that old ewe is an absolute thief, she is a hill sheep who comes in herself every year to the croft to have her lamb. Usually she just stays a day or two but this year she had twins and has not gone back to the hill. I don't know whether she thinks twins are better off in the croft or not, or is aware she has hardly a teeth left in her head and wishes to stay where the grass is not only greener but longer. I feed her every day, she is very fond of bits of bread. The funny thing is she stands on a dyke and looks down on my cabbages. She never eats them. Anyway she is very tame so if you like to catch, kill and bury her that is okay by me." Just at that time the Post Office door opened and in came the officer responsible for cruelty to animals. He said, "Mrs Hobbin, there is a dog chasing your sheep." Out quickly went my neighbour, then he came back to say 'sorry'. It was his dog and he hoped no harm had been done. I replied, "Don't worry, we all know how difficult it is to keep animals under control." I never heard the old moorit ewe mentioned again. She was still there when my husband came back from the whaling in May. He had to see to her demise himself.

The day after he arrived home, he said he supposed he would have to go and sharpen his tuskar. I was able to inform him I saw no need. I had had six men on flayed moor the week before and that, along with the ton of coal we usually ordered for the winter, should be enough. He really was quite surprised and relieved the job was done.

Spring was the time of year I liked very much, with the daffodils blooming and young lambs arriving. My husband and I took turns in getting up at 5am to see to the lambing. Fortunately we did not have a great many lambs so the season did not last long. When this was going on we went to bed early and one year I found it difficult to sleep. My husband said, "It is the magnetic field that is against you." I wondered what on earth he could mean. He explained that we had moved to another room for spring cleaning and now the head of our bed was not in the north. He would change the position of the bed and sleep would come. Sleep did come quicker but I honestly don't know if it was mind over matter or not.

The magnetic field, according to my husband, influenced other things as well. If one wanted to cut poans or diffets one should do it when the tide was flowing. Shaving the same, if you did not want to cut your face.

After the lambing it was time to prepare for the summer guests again, spring clean, paper and paint if required, and in between attend to the Post Office and shop. I was quite pleased at times to leave the spring cleaning to serve customers. It was nice to have a chat and hear what was going on in the outside world.

One elderly gentleman told me that he had been to Lerwick to sit a driving test for a moped bike that he had bought. He had not passed, even although he had explained to the examiner that all he needed it for was to go to the Brig Post Office for his pension. He had been unable to give hand signals. He said he could hardly let go of the handlebars long enough to touch the brow of his cap. The examiner said, "You can try again, even a dog is allowed a second bite. In the meantime you can drive if you put your L's up." With that a bit of a smeeg came over the examiner's face so the old gentleman just replied, "If you like I can put up the whole dashed alphabet." This elderly gentleman had been a sailor and had been all round the world. I just thought well he is not yet efficient at driving but how might the examiner get on steering a ship.

One diversion we had from time to time was people coming to the pier to pock sillocks. For those who don't know, a pock is a net in the form of a bag attached to a handle, for catching fish. A sillock is a coalfish in the first year. A few salt herring were usually used to sow them in. The herring were thrown over the pier into the water. In no time at all the sillocks would start to bool. Down in the water would go the pock and come up full of fish. We always shared in the fish, which we did enjoy. Quite often we were unable to eat them all so would boil the remainder for the hens. Unfortunately after some time the eggs started to taste fishy and the poultry would have then to survive on scraps and hens meal.

During one of those sessions, I heard about Auld Jamie's shortest story. Jamie and Johnnie were going to the pier to pock sillocks. Jamie said, "I will pock." "No", said Johnnie, "I will pock." "No", said Jamie, "I will pock and pock I pocket." So ended the story.

Life then carried on much as normal. Norman did not wish to go back to the whaling so any spare time he had was spent fishing for lobsters. During one of those trips he nearly lost his life. It had been a spell of very bad weather but one afternoon it fell away to a flat calm. Norman said, "Agnes, do you think I can go off to my lobster creels?" I replied, "Yes, it's a lovely afternoon for it." I thought it was the calm after the storm, in point of fact it was only a lull. He duly went and I plucked and cleaned two chickens, as I had a guest on a light diet. After that rather dirty work, I had a long soak in the bath plus a change of clothes. When I came out of the bath, I looked out the voe and could not believe my eyes. The waves were terrible and fish boxes were dancing around the pier. I knew no boat could make it in the voe, so I rang Gruting in the hope that Norman might have gone in there. No sign of him, but the gentleman who answered the phone said he would ring the coastguard for me. This he did and some time later the coastguard rang to say he had had a look round the shore in Walls but no sign of Norman or boat. Next the lifeboat people rang, the coastguard had been in touch with them. They were sorry but the weather was too bad for them to get out to their lifeboat. If the weather abated they would try.

A little while after this my son arrived home from school, carrying his sister. He talked about the dreadful weather and hoped his dad had secured the boat. I was obliged to tell him that not only was the boat missing, but so also was his dad. He wondered what I was going to do. I explained what had been done and now I was going to make our tea. After that I was going to cook late dinner for our guests. They always had dinner at night as they were usually trout fishing all day. "Well", said my son, "don't you think we ought to phone Maureen." This being my elder daughter now at school in Lerwick. "No", I said, "not yet, your dad is good at handling a boat, at the moment he is very fit and he has a lifejacket. He is no stranger to the sea. He survived thirteen days in an open boat after being torpedoed during the war, so I am still expecting him back."

Tea was prepared, served and eaten, as was the guests evening meal, with still no sign. By this time I felt sure I was a widow and felt it was my fault as I said he could go. People in the district began to ring and my head started to feel sore. After about another hour, the back door opened, and in came Norman with his gun over his shoulder and four lobsters in a bag. He took one look at me and said, "Agnes, did you not think to shut the garage doors in this terrible weather, we might have lost the roof." What an anti-climax. I explained that we thought we had lost much more than a garage roof — the coastguard had been out looking for him and the lifeboat was on stand-by for the weather to improve. It had been a terrible struggle to make the shore. He had looked up while hauling a lobster creel, saw the squall, dropped the creel and jumped from side to side of the boat to balance her. He knew he could never make it home to Bridge of Walls so he tried to manoeuvre the boat inside a place called the Green Head. This he managed with the help of the wind and beached the boat at a place called the Brunt Hill. With hindsight, he felt he should have walked down to Walls, it was nearer, and phoned to say he was on dry land, but he felt he should struggle for home. I now phoned the coastguard and lifeboat people who were jolly glad the emergency was over.

Time marched on, our two elder children went to college, Maureen to Aberdeen and George to Edinburgh. We were left with only Violet, who was still at school in Walls.

The summer came and guests came and departed. One angler borrowed our car to try fishing in a loch up north. When he came back he said he had had a strange experience. He asked someone to direct him to an inn where he could have a drink. On arrival there he ordered his drink but the proprietor did not seem keen to serve him. "You know", said the angler, "he even asked where I stayed and when I said Bridge of Walls House, he looked out the window. Then he said it is okay, I know that car, and I got my drink." I really had to laugh and informed him he had been drinking in a shebeen.

One day I was ready to take some sheep's puddings off the fire when the shop bell rang. I roared through, "I will be in a minute, when I have seen to my sheep's puddings." My customer said, "Are they home-made? I have not tasted one for ages. Do you sell them?" "No, but I can part with one." I got the tray through with the puddings and lifted one to put in greaseproof paper for her when another customer came in. Oh, mealy puddings, everybody wanted one. Well I could hardly give one to one and not to the others, so they all went. When my husband came in for his tea and sat down to scrambled eggs, he said, "I thought we would be having mealy puddings." I explained and said, "There will be no more till you get me some skins. When you do I hope you will not be like the old man and his dog."

The old gentleman was washing sheep's puddings in a burn. Every one he washed, he threw, without turning around, expecting it to land in a basin at his back. Unbeknown to him, his dog, who was also there, caught and ate each pudding before it reached the basin. After he had cleaned the last pudding he got up from his job, turned round and saw what had happened. He was so angry he intended to throw his tulley at the dog, in haste he made a mistake and threw the last pudding instead. The dog caught that one also and just glaepit it. I expect it was the only supper the dog got that night.

A couple and two teenage children all the way from America once had a holiday with us. They rather upset one of the girls serving the evening meal. They asked her if the natives were friendly. I said, "Somebody has been teasing them, pay no attention." Next day they came through to the kitchen, each carrying a big knife. They asked if it was safe to go out alone. I persuaded them they were perfectly safe, as long as they could swim. If they fell off the pier, that was our only hazard.

Another American stayed with us for a few weeks one summer. He rang up from Lerwick to say he had a problem. It was a public holiday and he was unable to get his travellers cheques cashed. Would it be possible, if I had a room, to take him on trust. Another thing, he was a vegetarian. I said that I would take him on trust and as long as he was just a vegetarian and not a vegan there was no problem. We had plenty of fish, eggs, milk and vegetables. What I did not tell him was that my husband bought for bait a box of small fish from time to time. It was absolutely fresh and I picked out the bigger ones for human consumption. That American was very appreciative of his food. He said he had never tasted such fresh salads nor had so many varieties. Another thing he liked was bread puddings. We required quite a lot of bread because the visitors were out all day with sandwiches. Because of this we often got a box of bread free gratis on a Saturday night. By Monday it would have been on the hard side to sell. From this box I would pick any currant bread to make bread puddings. Any stale buns or swiss rolls I made into trifles. The rest went to the hens.

The Christmas after he stayed with us, the American sent me three nice books. Unfortunately they were rather over my head, all about the ice age and other academic subjects. I passed them on to a more learned relative.

One winter when there seemed to be a little more time than usual, I suggested to Norman that he and I went to night class to learn machine knitting. This we did and were able to buy a knitting machine second-hand for £50. It had been decided that Norman would do the knitting and I the finishing. However, knitting his very first jumper he forgot to lay down loops for the oxters. I made a bit of fun, he got cross and said, "That is it, I will not knit any more." When I protested he said, "Agnes, I have papered and painted, put up plasterboard, made shelves, put them up, laid floors, done electrical work, made my own lobster creels, but this knitting lark is not masculine work. Let's sell the machine." This I would not agree to, so not only did I have to do the knitting myself, but also the Fair Isle and the finishing of the garment.

The first three jumpers I made took a lot of time. I had so often to pick up dropped stitches. However, when boarded and pressed they looked grand. I hung them on a pulley in the front kitchen, supposedly to dry them off, but really to admire them. I was not the only one taken with them. A travelling salesman came to the door with a pack. I was not going to take him in but he asked for a drink of water, so I felt obliged to say, "Come in and sit down while you drink it." After all, 'a cup of cold water to a stranger be given and great shall be thy reward in heaven'. Once the water was finished, he opened his pack and drew out a black nylon shirt. I bowed down to look at it and Norman said, "Black not required, I hope Agnes you are not going to wear away yet." The salesman held up his hand, "Now papa, you no speak, I do business with mama. I will give you the shirt for the three jumpers." This I refused and finally he offered three shirts for the jumpers. By this time Norman wanted the exchange. He thought I might have difficulty selling them. In point of fact, I sold them no bother and got an order for more.

At first I was very slow, but gradually I became faster and got regular orders. The only problem was I was always given enough wool for twelve jumpers. With stopping to serve in the Post Office, it was difficult to get them done in time. One afternoon Norman came in from the croft, said he was thinking he would have to stop his work and make his own afternoon tea. I informed him I was thinking the same thing, but had I stopped would have made him some along with mine. He then gave a good laugh and said, "Okay, I will make the tea and you shall have some with me."

In the end I had to sell the machine, Norman got quite cross about it. One day he said If I did not sell it, he would take it for a fasti for the boat. Then I asked him why he hated the machine so much. "Well", he said, "since it's arrival I feel as if there has been a bereavement in our home. After Violet has gone to bed there is just the two of us here. You continue to knit, there is no conversation, only yes or no if I ask you something." Well, what could I say, into the paper the machine went. When a week passed and there was no inquiry, I thought I was safe. Not so, at the end of the second week a lady rang up, came to have a look and said it was a good machine. After she had had a bit of a think she said she would offer £65 spot cash. I had no option but to sell, then it was I who felt there had been a bereavement. Knitting had been my interest for quite some time.

I was also disappointed as I had meant to concentrate on knitting and have a rest from summer visitors. After all, there were no overheads with knitting. No wages to pay out or cards to stamp, no linen or crockery to replace. Everything we seemed to make went back on the place. We now had a motor boat as well as the flat bottom. We had a reasonably good car, our first one cost £40. All second-hand beds had been replaced with new, except one iron bed that came from my old home. I refused to have it replaced as it was so comfortable. It had been procured for my mother a month or two before she died and it was possible to tighten the springs. When I left, years later, I wanted to take the bed with me but my son pleaded with me to leave it; he also liked it. Some time later, when I paid a social visit, I noticed the bed had gone and so asked my son what had happened to the bed. He explained that his wife wanted it replaced. He had refused, came home from his work one evening to find the bed gone, and another in its place. She refused to tell him where it had gone so he could not get it back. "Well", I said, "George, it is a bit like the story of the kitchen cupboard that the parrot chewed. As a newly married person I thought it was terrible and everybody else thought it was a fine cupboard."

In the end, we decided to let out a bit of the house that was self-contained for do-it-yourself summer visitors. Some of our old anglers did come back and were quite happy. I don't think the wives were quite so happy, but put up with it.

Our next change was when Norman went to work for a wholesaler in Lerwick. It worked well as he brought home our shop requirements himself. I was quite happy just being in charge of shop, Post Office, hens, dog and our home.

Life continued in this way for a number of years until a district nursing job came up in Walls. I applied, got it and transferred the Post Office back to my husband.

My gratuity for eighteen years as sub-postmistress was £440. I put it towards my first entirely new car, which cost at that time about £890. I enjoyed my new job, especially not being shut in the house most of the day.

# Chapter 5
# Back to Nursing

ONCE married I never expected to do any nursing again, but after a period of twenty-three years I returned to nursing. It came about this way. Walls and Sandness required a district nurse. I applied and was told I could get the job if I would go to Aberdeen and do the District Nursing Course. This I did and found a lot of changes had taken place in nursing. We now had disposable syringes and sterile dressing packs and all in all our work was much easier. We all had our own district but shared the work of any nurse who had a day off.

On one nurse's day off, the nursing officer looked around us all, and then said she would like me to give so and so her chair bath. My fellow nurses said, "Poor you, you have got her this time." It appeared she would only allow one nurse to wash her. Now this was a challenge and I made up my mind I would get the job done. On arrival I rang the bell, no answer so I rang again, then a cross voice said, "I am over eighty, it takes time." The door opened and an elderly lady said, "Who are you?" I replied, "Your relief nurse." Her reply to that was, "I only allow my own nurse to treat me." By this time I had my foot in the door, like a salesman, and asked if I might just come in for a minute as I was rather tired. This request was granted and I hurriedly got my coat off, turned it outside in so as not to carry germs from the outside world, and placed it on a chair, along with my nursing bag. With this movement my fob watch turned around, leaving only the steel side visible. "My", said my patient, "a nurse with a medal. I have never had a nurse with a medal before. I think I like the look of you. You may give me my bath." Thank goodness she never asked why I got the medal as I would have been hard pressed to answer. On completion of the job she said I could come again any time, a nurse with a medal would always be welcome in her home. Next day the nursing officer asked how I had got on and when I said, "No problem", my fellow nurses had a sly smile, all thinking it was a lie, until I enlightened them about the bogus medal.

One other time when I was helping on another nurse's day off, I had difficulty finding the proper road so asked directions from a taxi driver, waiting for his fare. Just then a priest appeared and the taxi driver had a quick word with him and the priest said, "Jump in nurse, I am in no hurry, I am just going to a funeral, we will give you a lift first." It turned out that the priest had a sister who did nursing back in Ireland and he and I had a good conversation on that journey. Next day the nurse I was relieving said she was sorry I had had to take a taxi to do her work. I replied, "No, I got a lift with a priest." Much laughter all round the table we nurses were sitting at, with a cry of "likely story".

My own district was Torry and I did enjoy it so much. The people were friendly and kind and I had many offers of tea or coffee. Another treat for me was to shop or window shop on my days off, and although when my course finished I was glad to go home to my family, I also felt a touch of sadness leaving my fellow workers and other friends that I had made.

However I soon got used to the quiet life again and settled down to look after the sick. There never seemed to be any shortage of people to be looked after so there was no time to get bored. In summer-time it was a pleasure to go to patients a long way off, but not so good when the snow and ice set in. I only went off the road once and it was not even snowing. It happened in this way. I took a short cut up a hill road in the dark and the car suddenly stuck in peat moor and would not move, either forward or backward. Fortunately I had a torch so was able to negotiate railings etc. and see my patient. Next I had to phone my husband and I thought he would think I was daft, but not so. He said, "Don't worry about it, get so and so to run you home. I will get ropes ready and we will have her out in no time." We did too, but I did not try anymore short cuts.

My husband was a great help with the car. He saw to the petrol, oil and water and always put her into the garage back first so she was ready for a quick getaway if I had a night call. Coming on for Christmas I always tried to visit the elderly, even if they did not require anything done. On one such occasion the gentleman of the house produced the whisky bottle. I explained that I did not drink and I had the car, but thank you just the same. However, the lady of the house said she would bring wine and she set before me a whole tumblerful, and a nice piece of cake. We had a nice bit of conversation and when I got up to go I thanked her for the cake and said how good the wine was. With this, she turned the bottle around and I saw, to my horror, that I had drunk at least three measures of sherry. Well I felt nothing wrong and proceeded home. When I got there I thought to myself, "I am every bit as good a driver as my husband, I will put the car in back first myself." It was no problem, she went in like magic. When I got into the house my husband said he would turn the car for me but I said, "Don't bother, I have done it myself." He looked at me and said,

"Agnes, you have had drink." "Well", I said, "I did have a glass of sherry, surely you don't grudge me that." His reply to that was, "Not at all, in fact, if one glass of sherry can do so much for your driving, I had better lay in a stock." I was more careful after that.

The next big change in my life was not a good one. My husband, who had had angina for some years, had a thrombosis and was taken to hospital. After two weeks he was recovering quite well, when he had another one and died. Now I had a decision to make. The Head Post Office offered me the chance to take the Sub-Post Office back in my name and I did think about it, but the pay was so little and I still had one child at school so I decided against it. The thing that worried me about the nursing was not the nursing itself but keeping the car in good shape and changing onto winter tyres etc. But I need not have worried, anybody and everybody was willing to help me. Once I stopped at the new Bridge of Walls Post Office and wondered if it would snow before I finished my visits, and me still with my summer tyres on. "No problem", said the sub-postmaster, "If you stand behind the counter a minute I will soon change the tyres."

Next my nursing officer gave me the chance to go on a two-week non-certificated Health Visitors' Course in St Andrews and I decided to accept it. I had a nice two weeks staying in the university students' halls of residence; good food, lectures and a bit of time to explore St Andrews. While there, the tutors from Dundee asked if anybody would like to do the full year's course and become a qualified health visitor. No-one wanted to do it, but I said if I had not been fifty-three years of age I would have done it. There followed a discussion on age and the tutor said there was no upper age limit, but they had turned away girls in their thirties who were too set in their ways and they knew would not adapt. When I came home I asked my nursing officer what she thought and it was decided I could go if I could get a college to accept me. I wrote to Dundee, because I had met two of the tutors at St Andrews, and I was asked to come to Dundee for an exam and an interview. This I did and almost immediately I got a starting date, but had to do four months at Ninewells Maternity Hospital first, so I could be better equipped to deal with post-natal cases. Accommodation was reserved for me in Maryfield Nurses Home and I arrived there one Sunday morning. It seemed to me quite dead and it took some time before I discovered my room. A day or two later when I was having my evening meal a lady in mufti sat down beside me, asked who I was, what I was doing and how I thought I might enjoy Maryfield. I explained that I thought nurses' homes had deteriorated badly. The room I had been given had badly scratched furniture, a dripping tap which I had to put a facecloth under to stop the noise, and a gap under the door of about two inches that allowed the smell of curry and rice to come through. Apart from that I had been informed I could get the students bus to Ninewells only to discover that it left at eight and I started at seven-fifteen, only on my lecture days could it be of any use to me. She seemed put out and two days later there was a note in my room to say that I had been allocated two rooms in Sister's quarters, one to sleep in and one to study in. The note also apologised for the room I had been given but the Shetland Health Board had not informed them that I was a qualified mature student. I later discovered that it was the nursing officer I had told about my room.

The next problem was the death of a sister-in-law in Shetland. I came back from Ninewells to find the telephone box locked up and I required to get through to my son about the funeral arrangements. When I ran the housekeeper to earth, she said she had locked it up because the nurses were using buttons and foreign money. She was still unwilling to open it, even for funeral arrangements, so I said I would have to have a word with the head postmaster, because this was not legal. What did I mean by that? I explained that I had been a sub-postmistress for eighteen years and had charge of four phone boxes. As a phone box was a public amenity, three days notice must be given before it was withdrawn and I was led to believe from fellow students that no such notice had been given. I got the box opened and it was never shut again in all the time I was there. Long after I left Maryfield Nurses Home, I discovered the Health Board paid a rent for the phone box, so the housekeeper could have refused to open it. Unfortunately by the time I found out it was much too late to apologise.

Life went on as normal, days in the wards and days at lectures, nights out with some of the other nurses, and finally it was over and I had a brief holiday in Shetland before starting at the college for the health visiting course. This was a course that I did enjoy in spite of all the new subjects like sociology, elements of medical genetics and psychology. I was lucky enough to rent rooms from a retired domestic science teacher. I had a bedroom and sitting room and shared her bathroom and kitchen. She was very set in her ways but I was old enough to understand her and young enough to adapt. I was only allowed two baths per week and except for washing out a few pants or tights, had to go to the laundrette. We shared the daily and Sunday papers and I cooked Sunday lunch for her and served it in my sitting room. She, in turn, treated me to Sunday afternoon tea in her drawing room. There I was able to admire her wonderful pictures, water-colours and paintings picked up at sales by her late father, when he was an auctioneer. I made many friends and was invited on my days off to other health visitor students' homes. In return, I served many a cup of tea or coffee in my sitting room as we often gathered there to swot together. I had my car brought down from Shetland and often took my landlady for a run on Sunday. We visited Scone Palace, Blair Castle and sometimes just went to the country to pick brambles.

As we approached the end of our year we all had to lecture. We had to choose a suitable subject related to our work in health visiting and our fellow students were our audience. They had to pretend to be whatever we wished them to be, for example, if we were lecturing about diet in pregnancy, they had to pretend to be pregnant. Also they were expected to ask questions and, if possible, catch the lecturer out. At one such lecture three of us decided to do follow-up lectures. One did the expectant mother, one did the birth and I did the care of the mother and baby when home from hospital. One student, who really liked to catch people out, asked if I thought it would be a good thing to put the white of an egg on a baby's sore bottom. I had never heard of this but thought if she was asking it must be okay, so I said, "Yes, I think it would be a good thing." She then followed it up by asking how long would the egg last. I had no idea, but quickly said it would all depend on the size of the egg and the size of the baby's bottom. Loud laughter in class, and I got a good report on my lecture. The sister tutor said that she did enjoy humour in lectures.

All too soon the year was almost over and we had a week of exams. When this was passed, one of my fellow students insisted that I came to her home for the weekend and quite a few of us had a meal out. Now this left only oral exams which were, of course, by external examiners. Next I heard that there was a seaman's strike on and the last day the P&O north boat to Shetland might sail was the day of our orals. I spoke to the sister tutor and explained that if I did not get my car home I would have no transport for my work and the strike could last for ages. She decided I could go immediately and get the car on the boat at Aberdeen, come back by train and she would put me in for the last oral exam. The car got a quick pack and my landlady was worried how I would manage as the weather was so bad. I explained that if I did not get lost, I would be okay. "Lost", she said, "you can't get lost, always keep the sea in sight." The journey was no problem and in no time at all I arrived at the boat, only to be told that the harbour bar was shut with the bad weather and I was to come back with the car tomorrow. In the end I explained my situation and the purser agreed to look after her. At that time the P&O ferry was further out and I thought I would not have time to walk to catch the train, so I went into the office to ask if they could phone a taxi for me. There was no phone available, the lines were down with the bad weather. I was just saying I would have to walk when an American came in and said he was having a taxi and he would take me. I explained about the phones and then thought where there is a steamer, there must be a steamer's store with people working, and it was just almost one o'clock. I duly went in to the store and asked if anybody was going home for their dinner and could they drop me at the railway station, and I would pay them. Almost everybody was willing and in no time the American (who had followed me in) and myself were at the station. I went into my bag for my purse but the American said, "I am paying", and he appeared to pay well. I got my train and arrived to find my landlady had the bath water hot and had even cooked me a meal. After a bath and a meal I got a taxi to the college for my oral exam which seemed nothing at all after the journey to Aberdeen and back on such a bad day. A few of my fellow workers were still waiting for their turn and we had a lot of fun when I told them about my journey, and one stayed behind to give me a lift home.

There was only a day or two of the course left and then I was home and doing full time health visiting. This was a job I really did like, particularly going to the babies when the midwife passed them over after the tenth day. The elderly had a special place in my life; I was getting older myself so could identify with them. Actually things were getting much better for most people in Shetland. North Sea oil was flowing, home helps were free and best of all, the Christmas bonus. When I became old enough to get and cash mine, I saw lots of smiling senior citizens also cashing theirs and thought if the powers that instigated this could see so many happy people, they would also feel rewarded. Sadly, I am now led to believe that this may not continue.

There was, of course, some trials on the district, the worst being dogs. I was bitten three times. It was usually my own fault because I had a habit of running if it was a bad day and this appeared to startle the dogs. The second thing was gates leading up to croft houses. They came in many shapes and rarities and sometimes I could not lift them, it was just a case of trailing them. There were times when I was tempted to leave them open if I was only going to be a short time with my patient, but I could not do it. Having been brought up on a croft I felt if a sheep got out, I would feel guilty. In spite of those trials, I felt as if health visiting was just made for me. My years in the Post Office stood me in good stead when I had to advise people about pensions and such like. One elderly lady was quite upset as she had recently given up her croft and said she could not live on the interest from her savings, which were too much for her to get Social Security. She had never paid National Insurance. I instructed her to get a form from the Post Office and request the old age pension, which was payable to everybody over eighty who did not have a pension, and was not means tested. She was over eighty and next time I came back I found a much happier patient.

My crofting background was also a help to me. Sometimes I helped to fill up crofting returns, especially if the one who usually did it was in hospital or had a stroke. It was also no problem to me if I found a person living alone who was too ill to go to the stack for peat. I just took the kishie, got the peat and lit the fire myself. Next I alerted the home help supervisor and soon services were in position.

During my time as health visitor a lot of newcomers came to Papa Stour so I had quite a few journeys there. Unless it was a very bad day, I quite enjoyed the sea journey and children old enough to walk used to come and meet me. One little boy took my hand and asked if I was very, very old. I said, "What age do you think?" He replied, "About ninety." His mother thought I would be offended, but I laughed and enjoyed the fun.

I retired when I was sixty-one and a half, and my fellow nurses gave me a lovely meal in the Westings Hotel, a camera, slide projector and screen. I was also treated to coffee mornings, afternoon tea and many gifts — the clock I received from the Young Weisdale Mothers still adorns my living room and keeps good time.

In all, my nursing days were enjoyable, if sometimes rather busy, and I would not have liked to miss them. Much has happened in the world since I started nursing in December 1940, I think and hope most for the better, and I would wish that all nurses starting today could have as full a life as I have had.

My story is now ending. All I can say in conclusion is that although life was very busy, it was never boring. However I would not like to work such long hours again. As one elderly lady said when a soldier asked to follow her home after a show dance, "No thank you, there is a season for everything."

# GLOSSARY

auld — old

bool — to jump through the surface of the water

cring — two lambs tethered together on one rope

diffets — big pieces of turf

fasti — a boats mooring rope

glaepit — swallowed greedily

gravit — a scarf or muffler

kishie — straw basket

lang — long

loops — stitches

misanter — sort of accident

moorit — brown

oxters — armpits

poans — thin flakes of turf

sam — same

sho — she

shocket — choked

siggy flowers — yellow irises

smeeg — a conceited smile

sow — tempt

trachled — overworked, put upon

trows — goblins

tulley — a large open knife with a wooden handle

wappit — thrown roughly